THE EXPOSURE ECONOMY

Reclaiming your Digital Identity in the Age of Surveillance Capitalism

Neil Singh

Published by Spines

ISBN: 979-8-90002-409-7

Spines

THE EXPOSURE ECONOMY

RECLAIMING YOUR DIGITAL IDENTITY IN THE AGE OF SURVEILLANCE CAPITALISM

NEIL SINGH

CONTENTS

Disclaimer vii
When They Stole My Mother's Dreams xiii
How to Use This Book xix
Foreword xxi

1. THE INVISIBLE CRISES 1
The Moment of Discovery 1
How the Attacker Knew Her 2
The First Wave of Fraud 2
The Emotional Cost 3
The Gradual Realization 4
The Larger Context 5

2. THE EXPOSURE ECONOMY EXPLAINED 7
What is the Exposure Economy? 8
How You Participate Without Choosing 9
Why It Persists: The Asymmetry Problem 10
The Surveillance Capitalism Connection 11
Why This Matters to Your Identity 12
The Five Layers of Your Digital Self 14
What You Can Do 16

3. THE $200 BILLION INDUSTRY BUILT ON YOU 17
(Data brokers, the invisible infrastructure) 17
How the Shadow Economy Actually Works 18
The Government Is Buying Too 19
Corporate Data Partnerships: Legal Identity Laundering 20
Why This Should Terrify You 21
The Bitter Tradeoff 22
How They Hunt You 23
The Real Cost 28
What Made Melissa Vulnerable 30
The Low-Income Vulnerability Gap 33

4. WHEN AI LEARNED TO WEAPONIZE PLAY 38
How Virtual Worlds Revealed Human Vulnerability 39
The Laboratory Scales: $2.9 Billion in Metaverse Losses 40
The $2.9 Billion Question 43
Youth Building Exploitable Identities in Real-Time 44
The Dual Mediation Reality: Why Security Theater Fails 45
The Bitter Truth About Beneficial AI 46
From Laboratory to Deployment 47

5. RECLAIMING YOUR DIGITAL IDENTITY 49
Section 1: What Reclamation Actually Means 49
Section 2: Individual Reclamation Strategies 53
Section 3: Why Individual Action Isn't Enough 63
Section 4: Psychological Reclamation 70
Section 5: What Happens Next 78

CONCLUSION 94

Practical Checklist 97
About the Author 103

DISCLAIMER

A Note on Privacy and Identity Protection

The individuals whose experiences are documented in this book, Sarah, Maria, Melissa, Robert, Lisa, and others, are real people who have endured genuine privacy violations, identity theft, data exploitation, and algorithmic manipulation. Their stories represent patterns documented in thousands of verified cases across regulatory filings, court records, consumer complaints, and investigative journalism.

However, to protect the privacy and security of those who have already suffered exposure, **all names have been changed and identifying details have been altered or consolidated**. In some cases, individual experiences have been

combined to illustrate common attack patterns while preventing re-identification of specific victims.

The case studies presented are **composite narratives** constructed from:

- Verified news reports and investigative journalism

- Federal Trade Commission complaint databases and settlement records
- Court filings in identity theft and data breach litigation
- Consumer protection agency case documentation
- Academic research on privacy violations and their impacts
- The author's professional research and consulting experience

While names and certain biographical details are pseudonymous, the **mechanics of the attacks, financial losses, emotional toll, and resolution timelines** accurately reflect documented reality. The data points, statistics, and institutional behaviors described are supported by cited sources throughout the text.

Platform and company breaches referenced in Chapter 4 (Roblox, Meta Quest, Decentraland, The Sandbox, and others) are documented incidents with verified reporting, cited with original news sources.

Academic research referenced throughout, including the Utilitarian-Hedonic Dual Mediation (UHDM) model *(authored by Neil and his two academic supervisors),* represents peer-reviewed published work cited in standard academic format.

This approach balances two competing obligations: first, to demonstrate that the threats described in this book are real, documented, and widespread; second, to avoid contributing to further exposure of individuals who have already been victimized by the systems this book critiques.

Surveillance Capitalism globally profits from treating people as data points. This book refuses to perpetuate that practice. The individuals behind these stories deserve both acknowledgment of their experiences and protection of their identities.

If you recognize elements of your own experience in these

pages, you are not alone. Millions face similar violations annually. The solutions presented in Chapter 5 are designed to help you reclaim agency regardless of where you are in your own journey through the Exposure Economy.

"The Exposure Economy systematically undermines human autonomy through AI-powered surveillance and manipulation. Reclaiming digital agency requires both individual privacy practices and collective advocacy for systemic reform."

DEDICATION

To my parents (father deceased, mom retired), whose challenges in life have shaped my purpose in ways they probably never imagined, without learning from them, my relentless advocacy for doing some good in the world within my profession of technology probably would not exist.

To my sister Amba Singh, who has fought the battles of difficulty and overcome immense hurdles we both had growing up, your fortitude throughout it all has made you a strong, wise human being. Thanks for fighting the good fight with me.
I dedicate this book to my son Eyan J Singh, who I hope understands one day that his father's goal was to make a genuine difference in our world for the good of humanity.

To the mentors and teachers (specifically Dr. Robert Davis and Dr. Bodo Lang) who have shaped my thinking and challenged me to grow 15 years ago, helping me realize my potential. Your wisdom and generosity have left an indelible mark on both my work and my life.

To the readers, past and future, who engage with these ideas: you are the reason this book exists. I hope it sparks thought, conversation, and action.

Finally, I acknowledge the countless unnamed individuals whose work, research, stories, and experiences have influenced my own. Though I cannot list you all, please know that your impact is felt and appreciated. I am profoundly grateful for the shared journey.

WHEN THEY STOLE MY MOTHER'S DREAMS

My mom flew 8,000+ miles from New Zealand to spend time with her grandson, my son. She'd saved for this trip, planned it for months, talked about all the things they'd do together. Instead, I watched her sit in my living room, staring at her phone while my little boy tried desperately to get her attention.

I thought she was just being rude. I had no idea she was drowning.

"Mom, can you put the phone away?" I asked more than once. She'd look up, smile apologetically, then within minutes be pulled back into that screen like it had some magnetic hold on her. My son would tug at her sleeve, show her his toys, and ask her to read a story. She'd nod absently, her fingers still scrolling, her mind somewhere else entirely.

This wasn't the mother I knew. This wasn't the woman who had raised me with fierce love and complete presence.

THE MOMENT EVERYTHING CHANGED

When she mentioned having trouble with her phone, I offered to take a look. Such a small gesture. Such enormous consequences.

I'll never forget the moment I opened her notifications. Thousands upon thousands of them. My stomach dropped as I scrolled through what looked like digital chaos emails from strangers, apps I'd never heard of, subscription confirmations for things that made no sense. Young men from countries I couldn't pronounce were sending my 65-year-old mother love messages. Cryptocurrency "experts" were promising her riches. Astrology sites were telling her fortune.

My hands were literally shaking as I realized what I was looking at. This wasn't just spam. This was an invasion. These people had colonized my mother's entire digital life, and she had no idea how to make it stop.

"Mom," I said, my voice barely steady, "how long has this been happening?"

The look in her eyes - confusion mixed with shame mixed with relief that someone finally knew that look will haunt me forever.

THE TERRIBLE TRUTH

What came next broke my heart in ways I'm still trying to understand. As we dug deeper into her accounts, her emails, and her banking records, the scope of what had happened to her became clear. For over two years, my mother had been living in a carefully constructed lie created by criminals who had studied her, learned her deepest vulnerabilities, and systematically destroyed her life.

They knew she worried about money. They knew she dreamed of leaving something behind for her children. They took those beautiful, human desires and weaponized them against her.

She had sent them nearly everything $300,000 she'd saved over thirty years of working as a modest career woman. Thirty years. Gone.

But here's what broke me most: she thought she was investing. She genuinely believed these "brokers" cared about her financial future. They had convinced her that secrecy was part of the process, that family members like me "wouldn't understand sophisticated investment strategies." They had turned her love for us into a reason to lie to us.

FIGHTING ALONE

When my sister and I started making calls to banks, lawyers, police, and anyone who would listen, we discovered something that made the whole nightmare even worse. Nobody cared.

The bank that had processed dozens of suspicious wire transfers to foreign accounts told us, "Your mother authorized these transactions." The police told us it was a "civil matter." The lawyers told us international fraud was "nearly impossible to prosecute." The authorities told us to file a report online.

Every single institution that should have protected my mother instead blamed her for not protecting herself.

I remember sitting in my car after one particularly devastating phone call with her bank, just crying. Not just for the money, though losing her entire life savings was catastrophic, but for the realization that we live in a world where predators can systematically destroy elderly people while the institutions we trust to protect society profit from looking the other way.

WATCHING HER BREAK

The financial loss was devastating, but watching what this did to my mother's spirit was infinitely worse. The shame consumed her. She couldn't understand how she'd been so "stupid." She couldn't forgive herself for hiding it from us. She couldn't accept that she'd been the victim of sophisticated criminals rather than her own poor judgment.

The woman who had raised me to be strong and independent was asking me questions like a lost child: "How did this happen to me?" "Why didn't I see it?" "What's wrong with me?"

Nothing was wrong with her. Everything was wrong with a system that allows this to happen.

THE RIPPLE EFFECT

This didn't just destroy my mother's retirement; it damaged our entire family. My sister, watching what technology had done to our mom, deleted every social media account she had. I found myself questioning every email, every phone call, every online interaction. Trust became a luxury we could no longer afford.

My son, who should have spent those weeks building memories with his grandmother, instead watched her struggle with shame and confusion. He was too young to understand why Grandma seemed so sad, why the adults were always whispering, why everyone seemed angry at people he'd never met.

The criminals who did this to my mother didn't just steal her money. They stole her dignity, her peace of mind, her relationship with technology, and precious time with her grandson that she'll never get back.

WHY I'M TELLING YOU THIS

I'm sharing our story because what happened to my mother is happening to someone's parent, grandparent, or loved one right now, today, as you read this. These aren't random acts of greed; they're part of a massive, sophisticated industry designed to identify and exploit the most human parts of us: our dreams, our fears, our desire to trust others.

The technology companies that make billions from our data aren't designing products to keep us safe; they're creating systems that make us more vulnerable. The financial institutions that process these fraudulent transactions aren't asking hard questions; they're collecting their fees. The authorities we expect to fight crime have decided that protecting elderly people from international fraud syndicates isn't worth their time.

My mother believed in a world where people could still be trusted, where opportunities were real, and where working hard and saving money meant security in old age. The criminals who targeted her knew exactly how to exploit those beautiful beliefs.

But here's what they didn't count on: they picked on someone whose children would fight for her. They stole from a family that refuses to stay quiet about what happened. They attacked someone whose story might prevent this from happening to your mother, your father, or your grandparents.

This is bigger than my family's tragedy. This is about the world we're creating and the people we're willing to sacrifice to digital predators. The question isn't whether someone you love will be targeted, it's whether we'll demand better protection before it's too late.

Every day we wait, more families are being destroyed. Every day we accept that "this is just how things are," more elderly people are losing everything they've worked for to criminals who face no consequences.

This is my mother's story. But it doesn't have to be yours.

HOW TO USE THIS BOOK

This book serves two distinct audiences facing the same challenge: protecting privacy and security when AI amplifies both threats and opportunities.

EMERGENCY PROTOCOLS LOCATION (FOR FAMILIES SEEKING IMMEDIATE HELP)

- For immediate crisis: Chapter 5
- For long-term protection: Start at the beginning of your relevant section and work through systematically.

PERSONAL PRIVACY & DIGITAL IDENTITY FOR: INDIVIDUALS, PARENTS, FAMILIES

You'll learn:

- What companies and criminals know about you
- How AI systems predict and manipulate your behavior

- Practical steps to reduce exposure immediately.
- Age-appropriate protection for your entire family

WHAT THIS BOOK IS

A research-based framework for understanding modern threats and taking meaningful action. A bridge between technical concepts and practical implementation. A roadmap for sustainable protection.

WHAT THIS BOOK IS NOT

A technical manual for experts. A legal guide (consult attorneys for compliance questions). A guarantee of perfect security (which doesn't exist). A one-size-fits-all solution.

FOREWORD

This book confronts what may be the defining challenge of our era: **protecting your privacy and security in a world where your digital identity has become a commodity**. The evidence presented in *The Exposure Economy* draws from multiple research methodologies - large-scale statistical analyses, consumer surveys, field experiments, and qualitative interviews - to provide the most comprehensive picture of digital privacy challenges and actionable solutions.

When different approaches converge on the same conclusions, you can trust the recommendations. Where evidence conflicts, I acknowledge uncertainty rather than overstate claims, giving you the honest assessment needed to make informed decisions about protecting yourself and your family.

This evidence-based approach combines academic rigor with practical application. After the final chapter, you'll find a comprehensive action checklist designed to help you take immediate steps to protect yourself and your family from identity theft and digital exploitation. The goal is not perfect privacy, that's impossible, but meaningful control over your digital life.

Authors' Personal Commitment to Readers: After reading *"The Exposure Economy,"* I genuinely hope that the public is much more aware of their vulnerabilities when they provide their personal data to any number of applications. The level of digital transformation dominating every part of our modern lives is unavoidable. All of us have a digital footprint, and to the best of my ability, I am trying to educate consumers about how to protect their digital identity by offering what I believe to be pragmatic information, which can give consumers immediate safeguards with their technology devices and applications.

CHAPTER 1
THE INVISIBLE CRISES
SARAH'S STORY: WHEN A DIGITAL SELF IS STOLEN

THE MOMENT OF DISCOVERY

It was supposed to be a quiet Tuesday morning. Sarah Johnson had just settled at her kitchen table, the aroma of coffee drifting from a steaming mug beside her laptop. Her phone buzzed, and she glanced down casually, then froze.

The message from her bank read: *"Large purchase detected $50,000 vehicle. Did you authorize?"*

Her hands trembled. The ceramic mug rattled against the saucer. "I thought maybe it was a glitch," Sarah recalled. "I don't even own a car that costs half that much. I just stared at the screen in disbelief."

She called the bank immediately, her heart pounding in her throat. On the other end, the fraud specialist confirmed her fear: someone else had been impersonating her. "We're seeing multiple attempts, Ms. Johnson," the representative said. "This isn't an isolated purchase."

Sarah described the moment as "like the floor fell out from

under me." In a single morning, her sense of financial stability and security was replaced with dread.

HOW THE ATTACKER KNEW HER

The fraud wasn't random. The attacker had carefully assembled Sarah's personal data from multiple breaches and sources.

- Her **Social Security number** was compromised during the massive **2019 Equifax breach**, which exposed data of over 147 million Americans.
- Her **mother's maiden name,** commonly used as a security question, had been revealed through a lighthearted Facebook quiz she took in college.
- Her **high school graduation year** was publicly listed on her LinkedIn profile.
- Even her **first pet's name**, often another "secret question," was available in an old Instagram throwback caption: *"Miss you, Max the best dog ever."*

"I couldn't believe it," Sarah said. "Every piece of information they used, I had given away years earlier without even thinking about it. Separately, it all felt harmless. Together, it was like they built a key to my entire life."

The intruder didn't need to "hack" in the Hollywood sense. They relied on

credential stuffing, **social engineering**, and information already circulating on the dark web from past breaches.

THE FIRST WAVE OF FRAUD

By the time Sarah realized what was happening, the damage was already accelerating. Within four hours of the first fraudulent purchase attempt, the attacker had:

- Opened **three new credit cards** in her name.
- Secured approval for a **$75,000 personal loan**.
- Tried to access her **401(k) retirement account**.
- Initiated steps to **refinance her mortgage**.

"Every time I thought I had shut one door, they were already walking through another," Sarah said.

The financial cost snowballed quickly. Between fraudulent withdrawals, legal fees, higher interest rates after her credit score plummeted, and lost income from missed work, Sarah estimated her total losses reached **$250,000**.

"I kept thinking, how is this even possible. I've worked my whole life, I've saved carefully, and in a matter of hours, someone else nearly erased all of it."

THE EMOTIONAL COST

The money was devastating, but Sarah emphasized that the emotional toll was harder to carry.

For weeks, she struggled with **insomnia**, waking in the middle of the night to recheck her bank apps. "I'd get up at 2 a.m. just to see if something new had happened," she explained. "My heart would race every time my phone buzzed."

She experienced **panic attacks** in everyday situations. "I'd be in the grocery store line, and I'd start sweating, feeling like I couldn't breathe. All I could think was, *what if something else is happening right now that I can't stop?*"

The breach didn't just take her money; it eroded her confidence in digital life. "I stopped shopping online. I didn't trust paying bills electronically. Even using my debit card at a gas station felt dangerous."

Work also suffered. The endless cycle of filing police reports, contacting credit bureaus, and documenting fraudulent charges consumed nearly **200 hours** of her time. She had to take unpaid leave to manage the crisis. "It wasn't just the money I lost; it was the hours, the stress, the energy. **It took over everything**."

THE GRADUAL REALIZATION

As she worked with investigators and financial institutions, Sarah slowly pieced together how her identity had been stolen.

- That *"fun quiz"* she took on Facebook had seeded answers to common security questions.
- The alumni form asking for her graduation year gave away a key data point.
- That old Instagram post about her dog had unlocked another layer of verification.

"Looking back, it felt like I had been leaving breadcrumbs for years. I just never imagined someone would pick them all up and use them against me."

This realization that her **digital self** had been exposed long before the attack was one of the hardest truths to accept.

THE LARGER CONTEXT

Sarah's case is far from unique. According to the **Federal Trade Commission (FTC)**, more than **1.1 million Americans reported identity theft in 2023**. Experts believe the true figure is even higher, since many cases go unreported.

Every **two seconds**, someone in the U.S. becomes a victim of identity theft. Losses are staggering: the **FBI's 2023 Internet Crime Report** estimated total identity theft losses at more than **$10 billion annually**.

Studies suggest preventive measures could make a difference. A recent analysis found that a **10% increase in people blocking app tracking** correlated with a **3.21% decrease in fraud complaints**. As Sarah later reflected, "If I had understood how much those small choices mattered, maybe I wouldn't be where I am today."

A CHANGED UNDERSTANDING

Six months after that first text message, Sarah's accounts were slowly stabilizing, but the scars remained. She described herself now as having **two identities**:

1. **Her physical self**, who lived her daily routines.
2. **Her digital self**, scattered across databases, apps, and platforms vulnerable, exposed, and largely outside of her control.

"My digital self feels like it's out there living a life I can't see," she said. "And someone else almost stole it completely."

The experience reshaped how she approached the internet. She became diligent about using password managers, enabling multi-factor authentication, and refusing unnecessary online surveys.

But she also carried a lasting skepticism. "I can protect myself better now, but I'll never feel the same way about logging in or sharing anything online."

THE UNANSWERED QUESTION

Sarah's story demonstrates that identity theft is not the result of a single mistake. It emerges from years of data exposure, systemic weaknesses in security, and the growing market for stolen digital identities.

For Sarah, the cost was nearly **$250,000**, months of disruption, and lasting psychological strain. For millions of others each year, the numbers may differ, but the story is similar.

As she put it, *"I always thought of identity theft as someone stealing a credit card number. I never understood it could mean stealing me."*

Her experience leaves us with a sobering question:

If someone can so easily steal your digital self, how much of your life is still truly your own?

CHAPTER 2
THE EXPOSURE ECONOMY EXPLAINED

What Sarah discovered too late was that her stolen identity wasn't just the work of one clever criminal; it was the byproduct of a vast, invisible economy that profits from knowing people more intimately than they know themselves.

When the fraudster pieced together Sarah's Social Security number, her mother's maiden name, her graduation year, and even her first pet's name, those fragments didn't come out of thin air. They came from years of seemingly harmless online activity, social media posts, casual quizzes, and account sign-ups that had been collected, stored, sold, and resold by companies she had never even heard of.

Sarah put it simply: "It felt like someone had been following me around for years with a notebook, writing down everything about me. Only I had no idea it was happening."

This is the "Exposure Economy."

Sarah's nightmare didn't begin when criminals drained her bank account. It began years earlier, every time she clicked "Accept" on terms of service she didn't read, every time she took a personality quiz for fun, every time she checked in at a restaurant or shared a photo with friends.

She was participating in what scholars now call the **Exposure Economy,** a global system where your identity becomes raw material, your behavior becomes product, and your privacy becomes the price of admission to modern life.

WHAT IS THE EXPOSURE ECONOMY?

The Exposure Economy operates on a simple premise: constant visibility creates value. Every digital action you take, every search, purchase, location check-in, social media post, and app download generates data. That data gets collected, aggregated, analyzed, and sold. The more exposed you are, the more valuable you become to the system.

But this isn't just about data collection. Traditional economies exchange goods and services. The Exposure Economy exchanges *you,* or more precisely, increasingly accurate predictions about your future behavior.

Companies don't want to know what you bought yesterday. They want to predict what you'll buy tomorrow, who you'll vote for next year, whether you'll get sick, change jobs, get divorced, or default on a loan. Your exposure to the accumulated record of your digital life makes those predictions possible.

The economy has three layers:

First, the infrastructure layer: Data brokers, advertisers, analytics firms, and technology platforms that capture, process, and trade behavioral data. This is the

$200 billion industry operating mostly invisibly, processing information on over 250 million Americans through companies most people have never heard of.

Second, the application layer: Insurance companies adjust your premiums based on lifestyle data. Retailers are changing prices based on your browsing history. Employers are screening candidates through algorithmic personality assessments.

Political campaigns micro-target messages designed to manipulate your specific vulnerabilities. Each applies predictions derived from your exposure to decisions that affect your life.

Third, the exploitation layer: Criminals, scammers, and identity thieves who weaponize the same data streams. They don't need to hack the system; they just need to access the exposure you've already created. Sarah's attackers didn't break into secure databases. They harvested fragments she'd scattered across a decade of normal internet use.

HOW YOU PARTICIPATE WITHOUT CHOOSING

The Exposure Economy's defining feature is involuntary participation. You can't opt out without opting out of modern economic and social life entirely.

Want a job? Employers expect you to have a LinkedIn profile and will Google your name. Want to stay connected with family? You need social media accounts that require trading your data for access. Want to compare prices, read reviews, or find basic information? You must accept tracking in exchange for free services.

The choice isn't between exposure and privacy. It's between exposure and exclusion.

This makes the Exposure Economy fundamentally different from previous economic systems. In traditional markets, you choose whether to participate in each transaction. You could decline to buy something, refuse to share information, or walk away from a deal you didn't like.

In the Exposure Economy, the transaction happens whether you agree or not. Facebook builds "shadow profiles" on people who've never created accounts, using data from users who uploaded their contacts. Data brokers create files on you from public records, purchase histories, and information scraped from websites you've never visited. Credit bureaus compile detailed financial profiles without your

Consent, you don't choose to participate in Equifax; you discover you've been participating when they lose your data in a breach.

Sarah didn't decide to expose her mother's maiden name, her childhood dog's name, or her high school graduation year. She mentioned them casually in social media posts over the years, creating puzzle pieces that criminals later assembled into skeleton keys for her security questions.

WHY IT PERSISTS: THE ASYMMETRY PROBLEM

The Exposure Economy thrives on information asymmetry. Companies know everything about you. You know almost nothing about them.

Data brokers track your behavior across hundreds of websites, but you can't see your own file without submitting written requests that 43% of brokers simply ignore. Algorithms make decisions about your creditworthiness, employability, and insurance rates, but the models are proprietary secrets you're not allowed to examine. AI systems predict your behavior with startling accuracy, but you can't access the predictions being sold to companies making decisions about your life.

This asymmetry isn't accidental. **It's the business model.**

If you understood exactly how much companies knew about you, you might object. If you could see the predictions they were making, you might correct errors or challenge assumptions. If you knew which data points led to being denied a job or charged higher prices, you might change your behavior or demand accountability.

The system operates in deliberate obscurity. Terms of service are written to be incomprehensible. Privacy policies describe data collection in vague language that obscures actual practice. When pressed, companies claim data is "anonymized."

Legal fiction researchers have repeatedly proven false. Eighty-seven percent of Americans can be uniquely identified from just three data points: birth date, gender, and zip code. Most "anonymized" datasets contain dozens more.

The Exposure Economy also persists because costs and benefits flow in opposite directions. Companies capture the value better through ad targeting, refined pricing strategies, and reduced uncertainty. You bear the risks of data breaches, discrimination, manipulation, and identity theft.

When Equifax lost 147 million Americans' personal information in 2017, the company paid $700 million in settlements. Sarah was in that breach. She received $6.50. The costs of exposure fall on individuals who can least afford them, while profits concentrate in corporations that externalize every risk.

THE SURVEILLANCE CAPITALISM CONNECTION

The Exposure Economy is the infrastructure. Surveillance capitalism is the ideology.

Scholar Shoshana Zuboff identified surveillance capitalism as a new economic order that claims private human experience as free raw material for behavioral prediction and modification. It doesn't just want to know what you'll do; it wants to shape what you'll do, guiding your behavior toward outcomes that serve corporate interests.

The distinction matters. Traditional capitalism sold you products. Surveillance capitalism sells *predictions about you* to third parties who want to influence your behavior. You're not the customer buying the product. You're the raw material being refined into predictions that get sold to actual customers, insurers, advertisers, employers, and political campaigns.

This transforms the relationship between companies and people. Maria's insurance company didn't just use her data to assess risk; they purchased a lifestyle analysis specifically to find justification for raising her premiums. The algorithm wasn't serving Maria.

It was serving the company's profit motive by finding patterns that statistically correlated with higher costs, regardless of whether those patterns applied to Maria individually.

Surveillance capitalism operates through what Zuboff calls "behavioral surplus" data extracted beyond what's needed to provide a service. Google doesn't need to track your location history to give you directions. Facebook doesn't need to analyze your photos for facial recognition to let you share memories with friends. Netflix doesn't need detailed viewing patterns down to when you pause or rewind to recommend shows.

But that surplus data, the excess beyond service provision, is where the real value lies. It enables prediction. And prediction enables control.

WHY THIS MATTERS TO YOUR IDENTITY

The Exposure Economy doesn't just collect data about you. It constructs a digital identity that increasingly determines how institutions see you and eventually, how you see yourself.

That digital identity exists independently of your actual self. It's assembled from fragments, interpreted through algorithms, categorized by systems that don't know you and weren't designed to understand human complexity. Yet this constructed identity makes real decisions about your life.

It determines what job postings you see, what prices you're offered, what content fills your social media feed, and which neighborhoods' real estate listings appear in your search results. The digital you assembled from exposure, refined through algorithmic analysis, shapes the opportunities available to the actual you.

Sarah discovered this when criminals used her digital identity against her. They didn't need to know the real Sarah, her values, her judgment, her actual security practices. They only needed the exposed Sarah: quiz answers, social media posts, public records, and data broker files. That fragmented, incomplete digital identity was enough to convince banks and retailers that the impostor was authentic.

This is the Exposure Economy's deepest threat. It doesn't just make you vulnerable to criminals. It replaces your actual identity with an algorithmic approximation, then treats that approximation as more authoritative than your own self-knowledge.

When Maria's insurance increased due to "lifestyle analysis," the company didn't ask about her actual lifestyle. The algorithm's interpretation of her data mattered more than her reality. When Melissa's credit score crashed because of fraudulent

accounts, institutions believed the digital record over her insistence that she hadn't opened those accounts. The system trusts the exposure more than it trusts you.

Reclaiming your identity in this economy isn't just about protecting data. It's about refusing to let algorithmic approximations define who you are and what you're allowed to become.

Sarah's exposure had been accumulating for years before criminals weaponized it. Every seemingly harmless digital interaction had added another piece to a profile she didn't know existed until that profile became a weapon aimed at her life.

Understanding how that profile was built is the first step toward preventing its construction in the first place.

THE FIVE LAYERS OF YOUR DIGITAL SELF

Sarah had assumed her "identity" was her name, Social Security number, and bank account. What she learned was that her digital identity was far broader, made up of at least five interconnected layers:

1. **Social Presence**

Every photo Sarah posted, every tag, every quiz she took became a data point. "That silly quiz about my mother's maiden name, who knew it would come back to haunt me?" she said. Social media companies don't just store posts; they analyze them to map personality traits, political leanings, and even vulnerabilities.

2. **Shopping Behavior**

Sarah often browsed electronics online, abandoning carts when she hesitated. Retailers tracked these habits. "I never realized companies knew that I was more likely to splurge on Fridays after work," she reflected.

Retailers track not only what you buy, but why you buy.

3. **Physical Movements**

Like most people, Sarah carried her phone everywhere. Location data revealed her routines, favorite restaurants, and when she was away from home. This data, sold to advertisers and brokers, quietly profiled her socioeconomic status and habits.

4. **Communication Patterns**

Even her messaging habits left a trail. Who she talked to, how fast she replied, what topics engaged her; these patterns were as telling as the messages themselves. Voice assistants and smart speakers add another layer, analyzing tone and even health signals.

5. **Health and Behavioral Habits**

Fitness apps and search histories told more about her than she realized. "I had no idea my late-night searches about stress and sleep could be used to profile me as 'high-risk,'" she admitted. Insurance companies pay for this kind of information to assess premiums.

Together, these layers created a comprehensive portrait of Sarah's life. Yet she didn't own any of it. Her information was scattered across hundreds of companies, many of which never responded when she requested data removal. Recent research confirms her experience: only **51.5% of data brokers respond to privacy requests** as required, while **43% fail to respond at all**.

If Sarah Had Known...

"If I had known my information was being sold like that, I would have been more careful," Sarah reflected.

Had she understood how the exposure economy worked, she might have taken steps to reduce her footprint: avoiding online quizzes, limiting public social media posts, using stronger privacy settings, and regularly checking data broker opt-out options. Even small actions like blocking app tracking could have mattered.

Studies show that a **10% increase in people blocking app tracking corresponds with a 3.21% decrease in fraud complaints.**

WHAT YOU CAN DO

Sarah's story is a warning, but it's also a guide. While no one can erase themselves completely from the exposure economy, individuals can take steps to reduce risk:

- **Audit your digital presence**: Search for your name online and see what's public.
- **Limit personal disclosures**: Don't post details like pet names, birthdates, or family information on social media.
- **Block app tracking**: Use privacy settings on your phone and browser to limit data collection.
- **Opt out of data brokers**: Request removal from major brokers like Acxiom and LexisNexis.
- **Use multifactor authentication and password managers**: Strengthen security beyond simple usernames and passwords.

Sarah's ordeal cost her close to **$250,000** and months of anxiety. But her experience highlights a larger truth: our digital selves are constantly being traded, tracked, and profiled without our awareness.

As she put it, *"It wasn't just my money they stole. They stole the version of me I didn't even know I had."*

And the larger question remains:

If your digital identity can be built, sold, and stolen without your consent, how much control do you really have over your own life?

CHAPTER 3
THE $200 BILLION INDUSTRY BUILT ON YOU

(DATA BROKERS, THE INVISIBLE INFRASTRUCTURE)

Maria Garcia didn't know Acxiom existed until her lawyer said the name during a deposition.

She was suing her health insurance company for raising her premiums 15% without explanation. No claims. No accidents. No changes to her policy. Just a letter saying, "lifestyle analysis indicates elevated risk factors."

"What lifestyle analysis?" Maria had asked three customer service representatives. None would tell her.

Her lawyer found the answer in the insurance company's vendor contracts. Acxiom, a data broker Maria had never heard of, had sold her insurance company a "risk profile" built from her fitness app, grocery store purchases, social media activity, and the neighborhoods her phone visited.

The algorithm concluded Maria lived a "high-stress lifestyle" because she worked late hours, bought convenience foods, and her phone showed she commuted through a "statistically higher-risk zip code." Never mind that Maria ran five days a week, didn't smoke, and had perfect health markers.

The profile cost her insurance company $0.47. It costs Maria $180 annually in higher premiums.

"Who gave them permission to buy information about me?" Maria asked her lawyer.

"Nobody had to," he replied. "You're not their customer. You're their product." Maria's story reveals the architecture of modern **surveillance capitalism.**

The average person's data is sold to more than 600 companies every year. Most people remain unaware that this trafficking exists, let alone understand its implications for their financial security, personal autonomy, and future opportunities.

Behind the visible layer of apps and websites operates a $200 billion global industry powered by data broker companies with names like Acxiom, Epsilon, and LexisNexis that most Americans have never heard of but that know everything about them.

These companies don't need your permission. They don't notify you when they create your profile. They don't tell you when they sell it. And research shows that 43% simply ignore you when you ask them to delete it.

HOW THE SHADOW ECONOMY ACTUALLY WORKS

Data brokers collect fragments of your life from hundreds of sources:

Every "like" on social media. Every product you browse but don't buy. Every app you download. Every location your phone visits. Your voting registration. Your property records. Your magazine subscriptions. Your warranty registrations. Your grocery store loyalty card.

They combine these fragments into detailed profiles that know you better than most of your friends do.

The largest broker companies processing data on over 250 million Americans sort people into thousands of behavioral and demographic categories. Not just age and income, but psychological profiles: "Urban Renters," "High-Net-Worth Seniors," "Financially Challenged Families," "Health-Conscious Millennials," "Impulsive Buyers."

Those labels determine what you pay and what opportunities you see.

When you search for flights, algorithms check your profile. iPhone users often see higher prices than Android users. The assumption is that iPhone owners will pay more. When you apply for a loan, lenders buy your profile to assess risk factors your credit score doesn't reveal. When you browse job listings, some opportunities never appear because algorithms have already decided you don't fit the profile.

You're being sorted, priced, and judged by companies that profit from knowing you, companies you'll never interact with directly.

And they operate with almost no oversight.

Research analyzing data broker compliance with privacy laws found that less than half respond to consumer deletion requests within legal timeframes. 43% simply fail to respond at all.

Why would they? You're not their customer. The companies buying your data are their customers. You're just the raw material.

THE GOVERNMENT IS BUYING TOO

Law enforcement doesn't need warrants anymore, not when they can buy what they want from data brokers.

Immigration and Customs Enforcement purchases location data to track undocumented immigrants. Local police departments buy consumer data to predict where crimes might occur. The IRS has purchased location data from apps to track tax evaders.

This end-run around the Fourth Amendment has become routine. Courts have ruled that warrants are required when the government *compels* companies to turn over data.

But if the government simply *purchases* data that companies sell commercially, many judges have found no constitutional violation.

Your phone's location history, showing every place you've been for months or years, is for sale to the highest bidder. And sometimes, the highest bidder is the government agency that would otherwise need a warrant to track you.

CORPORATE DATA PARTNERSHIPS: LEGAL IDENTITY LAUNDERING

Major companies share your data through complex partnership networks that sound cooperative but function as surveillance exchanges.

Your bank might share spending patterns with retailers who want to know your income level. Your phone company sells location patterns to advertising networks that want to know where you shop. Your insurance company buys health app data to identify customers who might file claims.

The magic word that makes this legal: **"anonymized."**

Companies claim they remove your name before sharing data, making it impossible to identify you. This sounds protective. It's largely fiction.

A landmark study demonstrated that 87% of Americans can be uniquely identified using just three data points: gender, birth date, and zip code. Most "anonymized" datasets include all three and dozens more.

Combining multiple "anonymous" data sets makes identification trivial. Your "anonymous" web browsing history, when matched with your "anonymous" purchase history and "anonymous" location data, identifies you more precisely than your name would.

Data flows through this network constantly. The photo you posted in Ohio gets stored on servers in Ireland, processed by algorithms in Singapore, analyzed by companies in India, and sold to buyers in twelve countries all within hours.

Different nations have different privacy laws, ranging from Europe's strict General Data Protection Regulation to jurisdictions with essentially no protection. Your data flows wherever companies profit most, making enforcement nearly impossible.

If a European company violates GDPR, it can simply move operations to a country with weaker laws. If an American company wants to avoid California's Consumer Privacy Act, it structures operations to fall outside its scope.

This global arbitrage means your data receives the weakest protection available anywhere in the chain, regardless of where you live.

WHY THIS SHOULD TERRIFY YOU

The data economy creates three cascading threats to your life:

First, financial manipulation. Dynamic pricing algorithms adjust what you pay based on your profile. Hotels charge more when they detect you've already booked flights. Car insurance quotes vary based on the websites you visit before requesting a quote. Stores show different prices to different customers based on purchasing power profiles.

You're paying a "data tax" on purchases throughout your life simply because companies know enough about you to charge the maximum you'll tolerate.

Second, democratic subversion. Political campaigns buy voter profiles detailing your fears, values, and vulnerabilities. They micro-target messages designed to suppress your vote or inflame your anger. Foreign governments do the same, purchasing access to American social media data to identify which messages will divide communities and destabilize democracy.

Research confirms this isn't theoretical: voters exposed to personalized political advertisements measurably change their voting behavior. Elections increasingly go to whoever buys the most sophisticated data, not whoever has the best ideas.

Third, algorithmic discrimination. AI systems trained on biased historical data perpetuate that bias at scale. If past lending data reflects discrimination against certain communities, algorithms learn to continue that discrimination, often in ways that are invisible and legally defensible because "the algorithm decided."

Low-income Americans are twice as likely to have limited privacy protections. They rely on free services that harvest data. They live in more heavily surveilled neighborhoods. They're less likely to know how to protect themselves.

The people who can least afford discrimination face the most invasive surveillance and the most harmful algorithmic judgments.

THE BITTER TRADEOFF

This system does provide genuine benefits.

Netflix knows what you'll enjoy watching. Google Maps routes you around traffic. Medical AI detects cancers earlier than human doctors sometimes can. Fraud detection saves your bank account from criminals.

These benefits are real. They improve lives. They sometimes save lives.

But you're not choosing this tradeoff consciously. You're not negotiating the terms. You're presented with a fait accompli: give us everything or get nothing.

Studies of Europe's GDPR found that privacy regulations reduced consumer participation in online activities by 12.5% people chose privacy over personalization when given a real choice. At the same time, advertising costs rose as behavioral targeting became more difficult.

Companies passed those higher costs to consumers through increased prices while continuing to profit from data sales wherever possible.

You bear both sides of the cost, reduced convenience *and* higher prices while corporations retain profits from your data.

Maria eventually settled her lawsuit. The insurance company admitted no wrongdoing but agreed to remove the "lifestyle analysis" from her risk assessment. Her premiums returned to normal.

But her data is still out there. Acxiom still has her profile. Hundreds of other brokers still sell information about her daily life. And next time, she might not have a lawyer who knows what questions to ask.

The $200 billion industry was built on you! It isn't going anywhere. The question is whether you'll let it continue operating in the shadows, or whether you'll learn how it works well enough to fight back?

HOW THEY HUNT YOU

Melissa Hart filed her taxes on February 14, 2019. Valentine's Day. She remembers because she joked with her boyfriend about spending the most romantic day of the year doing paperwork.

Three days later, the IRS rejected her return.

Someone had already filed using her Social Security number. Someone had already claimed her $8,600 refund.

"There must be a mistake," Melissa told the IRS agent on the phone. "I'm looking at my W-2 right now. Nobody else could have this information."

The agent's response chilled her: "Ma'am, we're showing a

return filed in your name on February 9th. It was accepted, and the refund was direct-deposited to an account in Florida."

Melissa lived in Kansas City. She'd never been to Florida. She didn't know anyone in Florida.

"How is this possible?" she asked.

The agent's answer would unravel Melissa's life for the next eighteen months: "Your identity has been compromised."

Melissa's nightmare demonstrates how modern identity theft works.

Digital criminals don't break into houses. They don't pick locks or smash windows. They build weapons from the fragments of your life scattered across the internet, fragments you didn't know existed, sold by companies you've never heard of, combined using techniques you wouldn't recognize.

Then they use your own identity against you with surgical precision.

Fourteen million Americans experience what Melissa experienced every year. That's one victim every two seconds. But the mechanics of these attacks remain invisible to most people until they're already bleeding.

Understanding how you are hunted is the first step toward not becoming prey.

Phase 1: Information Gathering

The attack on Melissa didn't begin on February 9th when someone filed her taxes. It began three years earlier, in 2016, when Equifax lost her data in a breach that exposed 147 million Americans.

Melissa never knew she was in that breach. Equifax sent a notification to an old email address she hadn't used in years. By the time she learned about it during the IRS investigation, the criminals had already weaponized her information.

The Equifax breach gave them her Social Security number, date of birth, and address. But that was just the foundation. They needed more to convincingly impersonate her.

They found it for free.

Melissa's LinkedIn profile listed her employer and job title. Her Facebook profile was set to public because she used it to promote her side business selling handmade jewelry showed her maiden name, her mother's name, and photos from her high school reunion that tagged the year and location.

One quiz she'd taken, "What Should Your Retirement Age Be?" asked for her current age, annual income, and dream retirement location. She'd answered honestly: 34, $52,000, somewhere warm.

A "Throwback Thursday" post showed her childhood dog, Biscuit. Banks love asking for pet names as security questions.

The criminals didn't hack anything. They harvested. They collected the digital breadcrumbs Melissa had scattered across a decade of normal internet use. Then they combined those fragments with data bought from brokers for pennies, the same data Maria's insurance company had purchased legally.

Melissa had unknowingly built their weapon, piece by piece, quiz by quiz, post by post.

The data illegal market where criminals trade this information operates openly on forums and encrypted messaging apps. A "fullz" complete identity package with SSN, DOB, address, mother's maiden name, and financial information sells for $15 to $150, depending on credit score and how recently the data was verified.

Melissa's "fullz" cost less than her morning coffee.

Phase 2: Social Engineering

With Melissa's data in hand, the criminals didn't immediately file fraudulent tax returns. They tested.

In December 2018, two months before filing Melissa's taxes, they called her bank.

"Hi, this is Melissa Hart calling about my account. I'm traveling and my debit card isn't working."

The bank's fraud prevention system flagged the call. The representative asked security questions.

"Can you verify your mother's maiden name?"

"Rodriguez," the caller answered. Correct. It was on Facebook. "And the name of your first pet?"

"Biscuit." Also correct. Thank you, Throwback Thursday. "What's your current employer?"

The caller, named Melissa's school district, was pulled from LinkedIn. The representative, satisfied, offered to send a replacement card.

"Actually, I'm overseas for another month. Can you just unblock the card I have?"

The representative complied. The criminals now knew Melissa's bank security was beatable.

They hung up without making any withdrawals. They were mapping defenses, not robbing yet. That would come later.

This is social engineering: using human psychology and institutional trust against itself.

The criminals knew something most victims do not realize: bank employees, IRS agents, and credit card representatives want to help. They are trained to verify identity through questions. But if you have the answers (purchased for pennies or scraped from social media), you pass verification.

The system is not designed to defend against someone *who is you*, on paper.

Phase 3: Account Takeover

On February 9, 2019, the criminals filed Melissa's tax return.

They used her real SSN, real employer information, and real W-2 data (estimated based on her salary from LinkedIn and public teacher pay scales). They claimed legitimate deductions. They made it look exactly like Melissa would file.

The only difference: they changed the direct deposit account to one they controlled in Florida.

The IRS accepted the return. Eight days later, $8,600 landed in their account. But they weren't finished with Melissa.

Once criminals successfully compromise one identity, they multiply it.

Over the next three months, while Melissa was still fighting with the IRS, the criminals opened four credit cards in her name. Two personal loans totaling

$19,000. A department store account that immediately maxed out at $3,500.

Each time, they used Melissa's real information to pass verification. Each time, they had statements sent to an address they controlled (a mail drop they had rented using a fake ID made with Melissa's real data).

By the time Melissa discovered these accounts in April, the criminals had disappeared. The cards were maxed. The loans were in default. And Melissa's credit score had cratered from 732 to 492.

The financial damage: $31,100 in fraudulent accounts. The credit damage: 240-point score drop. The timeline: Four months from first filing to full

discovery.

And Melissa's nightmare was just beginning…

Phase 4: Identity Multiplication

Advanced criminals don't just steal your identity once. They sell it to other criminals, creating a cascading marketplace of fraud.

Melissa's stolen identity appeared on the dark web marketplace in May 2019, three months after the tax fraud. Price: $45.

Why sell information you could keep using? Because used identities get flagged. Banks freeze accounts. Credit bureaus add fraud alerts. The well runs dry.

But to a different type of criminal, medical identity thieves, employment fraudsters, and synthetic identity creators, a "warm" identity that's already flagged for financial fraud is still valuable for other purposes.

In Melissa's case, someone used her identity to receive medical treatment at an emergency room in Texas (she discovered this during the investigation). Someone else used it to apply for unemployment benefits in her name in California.

Each new crime made Melissa's identity theft more complex and expensive to resolve.

Tax fraud alone is relatively straightforward to fight. Add credit fraud, and it becomes complicated. Add medical identity theft, and you're dealing with HIPAA privacy laws that make it nearly impossible to discover what treatments were received in your name. Add unemployment fraud, and you're fighting state bureaucracies across the country.

The investigation Melissa thought would take weeks took eighteen months.

THE REAL COST

The criminals who stole Melissa's identity netted approximately $39,000 in direct theft. They've never been caught. Melissa never recovered a dollar.

But the money was the least of what they took.

Financial Impact:

- $8,600 stolen tax refund (never recovered)
- $4,200 in legal fees, credit monitoring, and fraud resolution services
- $31,100 in fraudulent accounts (eventually dismissed but required legal intervention)
- Unknown hours at $25/hour: 437 hours documented = $10,925 in lost time
- **Total: $54,725**

Credit Damage:

- Score dropped from 732 to 492
- Denied apartment lease (had to break current lease early, paid $1,800 penalty)
- Denied car loan (had to buy a used car at 18% interest instead of a new car at 4%)
- Over ten years: estimated $50,000+ in higher interest on all borrowing.

Professional Consequences:

- School district policy required administrative leave during investigation (unpaid)
- Background check flagged fraudulent accounts as "financial irregularity."
- Missed two months' salary: $8,667.
- Nearly lost teaching license (required character review hearing)

Emotional and Physical Toll:

- 79% of identity theft victims report anxiety symptoms. Melissa developed panic attacks.
- 33% experienced sleep problems. Melissa was prescribed a sleeping medication.
- 41% report prolonged anger. Melissa's relationship with her boyfriend deteriorated, eventually contributing to their breakup.

Timeline to Resolution:

- February 2019: Discovery
- April 2019: Found credit fraud.
- June 2019: Denied apartment.
- September 2019: Administrative leave
- August 2020: Finally resolved
- Present day: Credit is still recovering.

Total time consumed: 437 documented hours across eighteen months.

That's eleven full work weeks. Nearly three months of full-time employment spent proving she didn't steal from herself.

WHAT MADE MELISSA VULNERABLE

Melissa was not careless. She was not ignorant. She was normal.

She used the internet the way everyone does. She shared information the way everyone does. She trusted that security questions would secure her accounts, that privacy settings would protect her data, and that companies would safeguard what they collected.

The system failed her in four specific ways:

First: The Equifax breach. Melissa never chose to give Equifax her data. Credit bureaus collect your information whether you consent or not. When they lost it, she bore the consequences. Equifax paid $700 million in settlements. Melissa received

$6.50.

Second: Data broker sales. The additional information that criminals needed income estimates, employment details, and lifestyle patterns, came from legal data broker sales. The same industry Maria fought in Chapter 2.

Third: Social media oversharing. Melissa's public Facebook profile answered every security question banks use. She didn't know these posts were security vulnerabilities. Nobody told her that Throwback Thursday could lead to identity theft.

Fourth: System design. Banks, the IRS, and credit card companies are designed to trust people who have the right answers to verification questions. Once criminals had those answers, the system worked exactly as designed. It verified the impostor and approved the fraud.

Melissa did not fail. The infrastructure failed her.

She could have made different choices, keeping Facebook private, never taking quizzes, freezing her credit after the Equifax breach she didn't know she was in. But the burden of perfect vigilance shouldn't fall on individuals navigating a system deliberately designed to extract and monetize their information.

The criminals who destroyed Melissa's life didn't innovate. They didn't break through sophisticated defenses.

They walked through doors left open by an industry that profits from exposure and an infrastructure that prizes convenience over security.

The Family Angle: Robert's Story

Melissa's attack was surgical and financial. Robert Chen's attack was emotional and devastating in different ways.

Robert was 72 when "Susan" sent him a friend request on Facebook.

Her profile showed a woman in her mid-60s, attractive, smiling on a beach. They had twelve mutual friends (all fake accounts the scammers had created). She lived two towns over. She was widowed, like Robert had been for two years.

They started messaging. Susan was warm, funny, and attentive. She asked about Robert's day. She laughed at his jokes. After two months, they'd talked every day.

Robert's daughter, posting about her family vacation on Facebook, mentioned her dad was home alone for two weeks. "Make sure you call Grandpa while we're gone!"

The scammers were watching.

Three days into the daughter's vacation, Susan's messages changed. Her mother had fallen ill in the Philippines (where Susan was "visiting family"). She needed money for emergency surgery. Just $5,000. She was desperate.

Robert wired the money.

The "surgery" had complications. Another $3,000. Then $7,000. Then "Susan" needed to pay hospital bills before they'd release her mother. Another $12,000.

Over eight months, Robert sent $147,000 to "Susan," sixty percent of his retirement savings.

Simultaneously, Robert received a call from "Microsoft" warning that his computer had a virus. The "tech support" representative gained remote access to Robert's computer. Saw his bank accounts. Noted the balances.

The "Microsoft" calls continued. Each time, Robert paid $300-500 to "fix" problems that didn't exist. Each time, the scammers screenshotted his financial information.

They were mapping his assets while bleeding them dry.

Robert's daughter discovered the fraud when she noticed her father had moved from his apartment to her basement. He couldn't afford rent anymore.

The scammers had studied Robert's family's social media for months before "Susan" made contact. They knew:

- He was widowed and lonely (daughter's memorial posts)
- His daughter traveled frequently (vacation photos)
- He had grandchildren he doted on (tagged photos)
- He had money (property records are public; his daughter mentioned his "comfortable retirement")

They knew exactly which emotional buttons to push.

THE LOW-INCOME VULNERABILITY GAP

Both Melissa and Robert had some resources to fight back. Melissa could afford a lawyer. Robert's daughter had space in her home.

But research reveals a disturbing pattern: low-income Americans are twice as likely to experience identity theft and have far fewer resources to recover.

They can't afford lawyers. They can't afford credit monitoring. They can't take unpaid time off work to make calls to banks and credit bureaus. They can't absorb the credit score damage that makes everything, housing, transportation, and employment, more expensive and harder to access.

Identity theft perpetuates economic inequality.

Victims lose money they can't afford to lose. They spend years repairing damage instead of building financial security. Stress destroys jobs, relationships, and health.

And the criminals know this. They specifically target people they perceive as less sophisticated about technology: elderly people, immigrants, people without higher education, people in financial distress who might be desperate enough to fall for work-from-home scams or advance-fee fraud.

The victims who can least afford to be victimized are the most.

The Compound Effect

Each identity theft incident makes victims more vulnerable to future attacks.

Criminals share victim information with other criminals. "This person fell for a romance scam" gets noted in databases. "This person's identity was successfully used for tax fraud" marks them as a tested, working identity.

Once targeted, victims face repeated attacks for years.

Melissa received phishing emails referencing her tax fraud case, scammers pretending to be IRS investigators asking for personal information "to resolve your case." Robert got calls from "government agents" claiming Susan had been arrested and needed bail money.

The stress of constant vigilance exhausts victims.

Many eventually give up and accept ongoing fraud as "the cost of living digitally."

The psychological toll is immeasurable. Melissa describes feeling "digitally homeless," unable to trust any online interaction, any phone call, any email. She closed all social media accounts. She stopped online banking. She pays cash whenever possible.

Some victims develop what psychologists call "digital anxiety disorder." They avoid beneficial technology because of past trauma. This digital avoidance limits their personal and professional growth in an increasingly digital world.

Prevention Is Possible

Melissa's and Robert's stories are devastating. But they're also preventable.

The criminals who attacked them didn't use sophisticated hacking. They used information that was freely available or cheaply purchased. They exploited human trust and institutional design flaws.

Basic security measures make you a harder target:

Identity theft prevention is like home security. You don't need perfect protection. You just need better protection than the next target. Criminals are rational actors seeking maximum return for minimum effort. They'll move to easier victims rather than work harder to attack you.

If Melissa had:

- Frozen her credit after the Equifax breach (free since 2018)
- Made her Facebook profile private
- Never answered quiz questions with real security answer information
- Used a password manager with unique passwords for each account

...the attack would have failed or moved to an easier target.

If Robert's daughter had:

- Not posted that he was home alone
- Taught him verification procedures for unexpected contact
- Set up spending alerts on his accounts
- Helped him understand romance scam patterns

...Robert would likely still have his savings.

The next chapters will show you exactly how to implement these protections. Not as overwhelming security theater, but as practical steps that take minutes to implement and genuinely reduce your risk.

Because understanding how you're hunted is only useful if you know how to not be prey.

- Sarah, from Chapter 1, lost $50,000 in a single attempted transaction that was caught.
- Maria, from Chapter 2, pays $180 extra every year because the legal data industry sold her profile.
- Melissa lost eighteen months of her life, her credit, nearly her career, and her relationship.
- Robert lost his retirement, his independence, and his ability to trust.

They are four people. There are fourteen million victims like them every year.

The criminals using these techniques face almost no consequences. The industry enabling them operates legally and profitably.

The infrastructure designed to protect you wasn't designed for digital-age threats.

But you're no longer walking blind into this system. You know how the attack works, phase by phase, piece by piece. Now you learn how to break the kill chain before it breaks you.

How do they protect themselves with their data being traded and compromised so abundantly?

"Melissa and Robert were attacked by human criminals who manually researched, crafted messages, and made phone calls. That was 2019. Today, artificial intelligence can do in seconds what took those criminals months. AI can analyze your social media, generate personalized phishing emails, create deep-fake audio of your family members, and predict which psychological vulnerabilities will make you respond. The four-phase attack hasn't changed. But the hunter is no longer human. It's a machine that never sleeps, never makes mistakes, and learns from every victim it claims."

CHAPTER 4
WHEN AI LEARNED TO WEAPONIZE PLAY

Lisa Chen stared at the ad on her phone during lunch break: "Expecting? Get ready with our pregnancy essentials starter kit."

She wasn't pregnant. At least, she didn't think so.

But the ad planted doubt. She'd been tired lately. Coffee tasted off. And yes, she'd browsed baby clothes online a few times just for fun, she told herself.

That afternoon, Lisa took a pregnancy test. It was positive.

The algorithm had detected her pregnancy three weeks before she did. By analyzing subtle shifts in her behavior, walking speed from her fitness tracker, coffee purchasing patterns from her credit card, and the timing of her late-night social media scrolling, the AI identified pregnancy indicators Lisa herself hadn't consciously noticed.

This represents a fundamental shift in power. AI systems can now know things about you before you know them yourself. And once they know, they use that knowledge to influence your choices in ways you may never detect.

But the algorithm that profiled Lisa didn't learn manipulation in a vacuum. It was learned in a laboratory that's been running for fifteen years, processing billions of behavioral experiments daily. A laboratory called gaming.

HOW VIRTUAL WORLDS REVEALED HUMAN VULNERABILITY

In 2013, researchers studying why people purchase and play video games uncovered something unexpected. The study, published in the academic journal *Internet Research*, analyzed 493 gamers across multiple game genres to understand what actually drove their behavior.

The conventional wisdom suggested people played games for utilitarian reasons to achieve objectives, mastered skills, and complete challenges. Game designers built around this assumption, creating complex progression systems, achievement tracking, and skill-based rewards.

The research revealed the opposite. Hedonic consumption enjoyment, self-concept, self-efficacy, and self-congruity drove game purchase and usage far more powerfully than utilitarian factors. Even more striking: utilitarian goals only influence behavior indirectly, flowing *through* hedonic experience rather than operating independently.

Translation: Players didn't buy games to accomplish tasks. They bought games to *feel* a certain way about themselves. The achievements were just the vehicle.

Gamers sought experiences that aligned with their self-concept, their perception of who they are and who they want to be. They pursued self-congruity, evaluating whether the game's image matched their ideal self. They needed self-efficacy; the belief they could master the game's challenges. And above all, they craved enjoyment as a subjective, emotional reward.

The practical implication for game developers: Stop optimizing for challenge complexity. Start optimizing for identity transformation and emotional resonance.

The gaming industry generated $24.75 billion in 2011 by implementing these insights. Players weren't paying for software; they were paying for permission to become someone else temporarily.

Nobody asked the obvious follow-up question: If virtual environments can exploit these psychological vulnerabilities to generate billions in revenue, what happens when AI applies these lessons to everything else?

THE LABORATORY SCALES: $2.9 BILLION IN METAVERSE LOSSES

By 2024, the answer had become devastatingly clear. Gaming and metaverse platforms lost $2.9 billion across targeted attacks exploiting the same psychological mechanisms that researchers had identified a decade earlier. But these weren't traditional hacks breaking through technical defenses. They were social engineering attacks that bypassed security by targeting identity, belonging, and fear of missing out.

The losses exceeded $2.9 billion across DeFi, CeFi, gaming, and Metaverse platforms, with access control vulnerabilities responsible for $1.7 billion in losses. "Access control vulnerabilities" is technical language for a simple truth: the attackers didn't break the locks; they convinced victims to hand over the keys.

In September 2024, Decentraland's official Twitter account was compromised and used to push a fake MANA airdrop and phishing link. Users who clicked and connected their wallets risked losing funds.

Why did sophisticated crypto users, people who understood blockchain security, fall for an obvious scam? Because the attack exploited self-congruity. Victims wanted to be seen as early adopters in their community, as insiders who got exclusive access. The scammers didn't hack wallets. They hacked identity.

The Sandbox experienced an internal breach where attackers sent phishing emails to users. Separately, the CEO's Twitter account was hijacked in May 2023 to promote a fake airdrop scam, leading to phishing attempts and possible asset loss.

Pattern: compromise a trusted authority figure, exploit community trust, trigger FOMO through artificial scarcity.

The Bored Ape Yacht Club and Otherside metaverse communities experienced Discord and Instagram account compromises that propagated phishing links, resulting in theft of NFTs and cryptocurrency.

These communities explicitly built identity around ownership. BAYC holders didn't just own expensive JPEGs; they owned membership in an exclusive club that signaled wealth, taste, and insider status. The attacks exploited exactly that self-concept vulnerability.

Researchers demonstrated "inception attacks" on Meta Quest headsets in March 2024, showing that if an attacker shares a network and convinces a target into developer mode or installing a malicious app, they can present a fake, undetectable VR environment and intercept or alter user actions.

This represents the ultimate evolution: AI creating entirely false realities that users cannot distinguish from authentic experience. Not hacking your perception, *replacing* your perception.

The Pattern: From Loot Boxes to Life

The techniques gaming platforms developed to maximize engagement and spending have become the foundation for AI manipulation across every domain.

Loot box psychology → Dynamic pricing algorithms

Games discovered that variable reward schedules, the same mechanism that makes slot machines addictive, drove more spending than fixed prices. You didn't know if the next box would contain rare items, so you kept buying.

AI pricing systems now apply identical logic. Hotels charge more when algorithms detect you've already booked flights you're committed to, creating artificial scarcity that triggers compulsive purchasing. You know you should wait, compare prices, and think rationally. But the hedonic trigger overrides utilitarian planning.

Avatar identity formation → Digital twin vulnerabilities

Gamers created idealized avatars representing not who they are but who they wish to be. The research found significant value in the gap between actual self-concept and ideal self, the psychic distance players sought to temporarily close through gameplay.

Sarah (from Chapter 1) built a digital identity through quizzes and social media that didn't match her security-conscious real self. That gap between her careful offline behavior and her casual online sharing created the vulnerability that criminals exploited.

Social proof in MMORPGs → Instagram FOMO

Multiplayer games taught AI how peer pressure drives spending. Players bought cosmetic items not for functionality but for status to impress others or avoid ridicule. The game environment provided a platform for self-image value, allowing consumers to gain worth from assuming simulated characters and transforming their self-concept temporarily.

Instagram's AI learned the same lesson. It doesn't show you the products you need. It shows you products your social circle values, triggering comparison-driven purchasing through carefully curated feeds designed to highlight gaps between your life and your idealized self.

Perceived enjoyment as an override mechanism

The 2013 research identified enjoyment as a self-oriented reward associated with escapism. It served as the primary hedonic determinant, often overriding rational utilitarian considerations about cost, time, or practical value.

Dating apps manipulate enjoyment by timing notifications for moments when you're lonely or vulnerable, emotional windows identified through behavioral tracking. The apps might hide compatible matches to create frustration that drives premium purchases. The goal isn't your happiness. It's maximizing your engagement and spending by exploiting the hedonic need for connection.

THE $2.9 BILLION QUESTION

If gaming taught AI how to manipulate hedonic drivers at scale, what's the cost when these techniques deploy beyond entertainment?

- Maria (Chapter 2) pays $180 extra annually because algorithms profile her "lifestyle risk" by analyzing behavioral patterns identical to those studied in gaming research, except the game is her insurance premium.
- Melissa (Chapter 3) lost eighteen months fighting identity theft that succeeded because criminals exploited the same self-concept vulnerabilities. They didn't just steal her data. They impersonated her convincingly enough that institutions designed to protect her authenticated the impostor.
- Lisa's pregnancy algorithm didn't innovate a new manipulation. It applied gaming industry insights about behavioral prediction and hedonic trigger timing to medical advertising.

The laboratory experiment worked. Too well.

YOUTH BUILDING EXPLOITABLE IDENTITIES IN REAL-TIME

Forty-seven million children under thirteen use Roblox daily. They're not just playing games. They're building persistent digital identities that will follow them for decades.

Every purchase, social interaction, behavioral pattern, and identity experimentation gets recorded and analyzed. Account takeovers surged 1,000 percent, with many targeting gaming accounts where users had built valuable digital assets and identities.

AI systems analyze this data to build predictive models of who these children will become as adults, models that will influence their college admissions, job opportunities, insurance rates, and loan approvals years before they're old enough to consent to data collection.

Throughout 2023-2024, many metaverse projects and communities fell victim to phishing campaigns masquerading such as token airdrops, wallet migrations, or mints. Victims who connected wallets or signed malicious transactions lost NFTs and cryptocurrency. These attacks succeeded by exploiting the same psychological vulnerabilities, trust, FOMO, and desire for exclusive access that drive in-game purchases.

The pattern repeats: Create artificial scarcity. Trigger social proof. Exploit self-congruity. Bypass rational processing through hedonic appeal.

Except that children's developing brains are more vulnerable to these manipulations than adults. They're building identity formation patterns that normalize constant surveillance, behavioral profiling, and psychological manipulation as the cost of digital participation.

THE DUAL MEDIATION REALITY: WHY SECURITY THEATER FAILS

The 2013 research tested two competing models: one where hedonic and utilitarian factors independently influenced behavior, and one where utilitarian goals operated indirectly through hedonic experience. The data strongly supported the dual mediation model; utilitarian concerns only mattered when filtered through emotional resonance.

This explains why traditional security fails against AI-driven threats.

Banks, credit bureaus, and identity verification systems operate on utilitarian assumptions. They believe people will protect themselves through rational cost-benefit analysis: "If I use a strong password, I reduce fraud risk."

But AI attacks exploit hedonic vulnerabilities. The phishing email doesn't ask you to "compromise your security for no reason." It triggers FOMO ("exclusive offer expires in 24 hours"), social proof ("your colleagues have already claimed this"), or self-concept alignment ("as a valued premium member, you're invited...").

Your utilitarian knowledge that you *should* verify the sender becomes irrelevant when hedonic triggers activate.

Lisa didn't ignore the possibility of pregnancy because she was stupid. She ignored it because the pregnancy ad bypassed rational processing entirely, triggering doubt through emotional resonance before her analytical mind could engage.

This is why security awareness training shows minimal effectiveness. You're trying to strengthen utilitarian decision-making in a system where hedonic factors override rational analysis.

THE BITTER TRUTH ABOUT BENEFICIAL AI

AI-driven personalization provides genuine benefits. Netflix knows what you'll enjoy watching. Google Maps routes you around traffic. Medical AI detects cancers earlier than human doctors sometimes can.

These improvements are real. They save time. They sometimes save lives.

But the same systems that help you also manipulate you, and they learned *how* to manipulate you in virtual environments specifically designed to bypass rational resistance.

The research concluded that while utilitarian consumption has value, hedonic rather than utilitarian consumption positively impacts purchase and usage. Both exist, but hedonic experience mediates utilitarian goals rather than operating separately.

AI learned this lesson perfectly. It doesn't ask permission to manipulate your emotions. It doesn't explain that the convenience you receive costs you autonomy, privacy, and freedom from behavioral exploitation.

It simply optimizes for the metrics it was trained on, and those metrics, inherited from gaming, prioritize engagement and spending over well-being.

The algorithm serving Lisa's pregnancy ad was genuinely trying to be helpful. Targeted advertising reduces irrelevant noise, showing you products you actually might need based on behavioral analysis.

But "helpful" and "invasive" aren't opposites. They're the same capability viewed from different angles.

The AI that helpfully identifies your pregnancy three weeks early is the same AI that allows insurance companies to price-discriminate before you've disclosed medical conditions. The prediction model that improves your experience also enables exploitation.

FROM LABORATORY TO DEPLOYMENT

In 2013, gaming research revealed how virtual environments manipulate human psychology for profit. The findings were academically interesting but seemed contained to entertainment.

By 2024, criminals were exploiting these exact patterns but at an AI-enabled scale and across every domain of digital life.

The Decentraland hack, the Sandbox breach, the Roblox account takeovers, together with hundreds of other incidents totaling over $2.3 billion lost across 760 on-chain security breaches, weren't flukes. They were proof of concept.

AI had industrialized the manipulation techniques that gaming had pioneered. Virtual worlds taught machines how to exploit self-concept, self-congruity, and hedonic override. Now those lessons apply to your healthcare (Lisa), your insurance (Maria), your identity (Melissa), your retirement (Robert), and your children's developing sense of self.

The laboratory experiment ended. The real-world deployment began. And unlike gaming, there's no reset button.

Melissa and Robert were attacked by humans who manually researched, crafted messages, and made phone calls. That was 2019.

Today, artificial intelligence can do in seconds what took those criminals months. AI can analyze your social media, generate personalized phishing emails, create deepfake audio of family members, and predict which psychological vulnerabilities will make you respond.

The four-phase attack from Chapter 3 hasn't changed. But the hunter is no longer human. It's a machine that never sleeps, never makes mistakes, and learns from every victim it claims. You can't fight this threat the same way you fight human attackers. The scale is different. The speed is different. The psychological precision is different.

But you're not defenseless. Understanding how AI learned to manipulate humans in virtual laboratories where millions of children still play is the first step toward protecting yourself in a world where the hunter operates at machine speed.

CHAPTER 5
RECLAIMING YOUR DIGITAL IDENTITY

SECTION 1: WHAT RECLAMATION ACTUALLY MEANS

Sarah eventually regained control of her accounts. The bank reversed most fraudulent charges. Her credit score recovered after two years. But something fundamentally changed. She describes feeling "digitally homeless," unable to trust any online interaction, any phone call, any email.

Maria's insurance premiums returned to normal after her settlement of the lawsuit. But her "lifestyle analysis" profile still exists, sold to hundreds of companies she'll never know about. Every year, new buyers access predictions about her health, her habits, and her future predictions she cannot see, challenge, or correct.

Melissa spent eighteen months proving she didn't steal from herself. She won. But the digital records of those fraudulent accounts still appear in background checks, requiring her to explain the same theft repeatedly to employers, landlords, and loan officers who wonder why someone innocent has so much documentation of fraud.

Robert moved in with his daughter. He's financially secure but emotionally devastated, unable to trust anyone online. He canceled his email account, deleted his social media, and refuses to answer phone calls from numbers he doesn't recognize. He reclaimed safety by abandoning digital life entirely.

Lisa had her baby. The algorithm continues predicting her behavior with disturbing accuracy, now using her purchasing patterns to build a profile on her daughter.

Before the child turned six months old, data brokers had assembled files tracking the family's movements, purchases, and behavioral patterns.

None of them fully reclaimed what was taken. That's the honest starting point for this chapter.

Understanding Reclamation vs. Protection

Most privacy advice focuses on prevention: stop future harm before it happens. Use strong passwords. Enable two-factor authentication. Don't click suspicious links. These strategies matter, but they're not reclamation.

Reclamation means recovering the agency you've already lost. Resisting ongoing extraction. Rebuilding identity on your terms rather than accepting algorithmic approximations. Taking back power in a system designed to concentrate it in corporate hands.

The distinction matters because prevention assumes you start from a position of safety. But after reading Chapters 1-4, you understand the truth: you've already been exposed. Data brokers already have your files. AI systems already predict your behavior. Your digital identity already exists, assembled from fragments you scattered across years of normal internet use.

You can't prevent what's already happened. You can only reclaim what's been taken or at least limit how much more gets extracted going forward.

The Hard Truth About Digital Reclamation

You cannot completely reclaim your digital identity. The system doesn't allow it.

Data already collected cannot be unlearned by AI systems. Even if a broker deletes your file, the machine learning models trained on that data retain the patterns. Your behavioral fingerprint persists in algorithmic weights and biases that no deletion request can reach.

Legal rights to deletion don't remove data from third-party systems that purchased it before your request. When Acxiom sells your profile to an insurance company, that company isn't bound by your deletion request to Acxiom. The data proliferates faster than you can chase it.

The Exposure Economy's infrastructure persists regardless of individual action. Your opting out doesn't slow the $200 billion industry. It doesn't change surveillance capitalism's business model. It doesn't weaken the systems that profit from exposure.

This sounds hopeless. It's not. It's realistic.

Understanding what you cannot change clarifies what you can:

You cannot eliminate your exposure. But you can limit further extraction.

Every privacy protection you implement reduces how much additional data gets collected. It doesn't erase the past, but it changes the future. Think of it like stopping a leak; the water already on the floor remains, but you prevent more damage.

You cannot delete algorithmic predictions. But you can exercise what rights exist.

Even partial enforcement of privacy laws creates friction for the industry. When 57% of data brokers eventually respond to deletion requests, that's 57% less data available for sale. It's insufficient, but it's not nothing.

You cannot opt out of the Exposure Economy. But you can build psychological resilience against algorithmic manipulation.

This is where your unique power lies. AI can predict your behavior, but prediction isn't control. Understanding how algorithms exploit hedonic vulnerabilities, your need for belonging, status, and self-concept alignment lets you recognize manipulation and choose different responses.

You cannot reclaim your identity alone. But you can join collective efforts for systemic change.

Individual protection creates modest personal benefit while leaving the exploitative system intact. Collective action through advocacy, litigation, and democratic pressure targets root causes rather than symptoms.

What Success Actually Looks Like

Reclamation isn't binary. You won't wake up one day with your digital identity fully restored, your privacy perfectly protected, the Exposure Economy dismantled.

Success means:

Reducing exploitation incrementally. Each data broker deletion request, each privacy setting adjustment, and each refusal to share unnecessary information decreases your exposure slightly. These small reductions compound over time.

Resisting psychological manipulation consciously. You'll still see targeted ads, receive personalized recommendations, and encounter algorithmic nudges. But you'll recognize them as manipulation attempts rather than authentic desires, making deliberate choices instead of reactive ones.

Contributing to collective power gradually. Supporting privacy advocacy organizations, participating in class actions, and voting for candidates who prioritize privacy protection, these actions feel small individually but create political pressure for systemic reform.

Living with technology on more equitable terms. Not

abandoning digital life like Robert, but not accepting total surveillance like the default demands. Finding sustainable practices that let you participate in modern society while maintaining boundaries that the system would prefer you didn't have.

Reclamation is harm reduction, not elimination. It's regaining agency within constraints, not transcending those constraints entirely. It's refusing to let the Exposure Economy's existence define the limits of your resistance.

The rest of this chapter shows you how.

SECTION 2: INDIVIDUAL RECLAMATION STRATEGIES

Individual action alone cannot solve systemic problems. But it can reduce your personal exposure, exercise existing rights, and make you a harder target than the millions who take no protective action.

These strategies won't make you invisible. They'll make you less profitable to exploit.

Step 1: Audit Your Exposure (Know What Exists)

You cannot reclaim what you don't know exists. Start by discovering what the Exposure Economy has already assembled about you.

Check Known Data Breaches

Visit haveibeenpwned.com and enter your email addresses. This free service, created by security researcher Troy Hunt, searches over 12 billion compromised accounts from documented breaches.

If your email appears, the site shows which breaches exposed it and what data was compromised: passwords, security questions, purchase history, and personal information. This doesn't show everything (many breaches go unreported), but it reveals confirmed exposures.

For each breach, change the password on that account immediately if you haven't already. Assume that any security questions using information from that breach have been compromised.

Request Your Data Broker Files

Data brokers must respond to access requests under various state laws, though compliance remains inconsistent. Request your files from the five largest brokers:

- **Acxiom**: Visit acxiom.com/about-acxiom/privacy/opt-out and submit a "Request My Data" form
- **Epsilon**: Email privacy@epsilon.com with subject line "Data Access Request"
- **Experian**: Use experian.com/privacy, but note this is the credit bureau; the marketing division has a separate request process
- **Oracle (formerly Datalogix)**: Visit oracle.com/legal/privacy/marketing- cloud-data-cloud-privacy-policy.html
- **LexisNexis**: Use optout.lexisnexis.com for some data; full access requires a written request

Reality check: Research shows 43% of brokers fail to respond to deletion requests. Access requests fare slightly better but still face significant non-compliance. You'll likely need to follow up multiple times. Document every request with dates and confirmation numbers.

When files arrive (if they arrive), you'll see categories you've been sorted into: "Urban Renters," "High Net Worth Seniors," "Financially Challenged Families," plus hundreds of behavioral and demographic markers. These categories determine what ads you see, what prices you pay, and what opportunities appear or don't.

Google Yourself Thoroughly

Search your full name in quotes, plus your city or employer if you have a common name. Review the first ten pages of results, not just the first page. Data brokers, people-search sites, and cached information often appear deeper in results.

Search variations: maiden names, nicknames, previous addresses, usernames you've used on forums or social media. Each variation reveals different fragments of your digital identity.

Use incognito mode or a different browser to see results without personalization. Your regular Google account shows you what Google thinks you want to see.

Incognito shows what others researching you will find.

Record what you discover: which people-search sites list your information, which social media profiles appear, which old forum posts or comments surface, which photos Google associates with your name. You'll use this inventory to systematically request removals.

Review Social Media Presence

Search your social media accounts (Facebook, Instagram, LinkedIn, Twitter/X, TikTok) for posts containing:

- Full address or specific location tags
- Phone numbers or email addresses
- Children's names, schools, or ages
- Financial information or employment details
- Travel plans or vacation announcements
- Photos revealing home interior, valuables, or security systems

These posts don't just expose you; they expose your family, revealing patterns criminals exploit. Robert's daughter's vacation announcement told scammers exactly when he was home alone and vulnerable.

Delete or edit posts that reveal too much. Change privacy settings so future posts default to "friends only" or more restrictive. Location sharing should be disabled entirely or limited to specific trusted contacts.

What This Audit Reveals

Most people discover they have:

- 5-15 documented breaches across various accounts
- Files at 3-5 major data brokers (those who respond)
- 20-50 people-search site listings
- Hundreds of social media posts containing personally identifiable information
- Multiple email addresses and usernames are tied to their identity

This exposure accumulated gradually, invisibly, over the years. Each piece seemed harmless individually. Assembled, they form a comprehensive profile that anyone with internet access can access, including the criminals who attacked Sarah, Maria, Melissa, and Robert.

Step 2: Exercise Legal Rights (Use What Power Exists)

Privacy laws remain weak and inconsistently enforced in the United States, but some rights do exist. Use them.

California Consumer Privacy Act (CCPA)

If you're a California resident, the CCPA grants you:

- Right to know what data companies collect about you
- Right to delete personal information (with exceptions)
- Right to opt out of data sales
- Right to non-discrimination for exercising these rights

Even if you don't live in California, many companies extend CCPA rights nationally rather than maintain separate systems for California users. Submit requests and see what happens.

General Data Protection Regulation (GDPR)

If you interact with companies offering services in the European Union, GDPR grants stronger protections:

- Right to access your data
- Right to deletion ("right to be forgotten")
- Right to data portability
- Right to object to processing

Many international companies (Google, Facebook, Microsoft) provide GDPR request forms usable by anyone, not just EU residents. The companies may deny non-EU requests, but some honor them to avoid operational complexity.

Credit Freeze (Free for All Americans)

Since 2018, credit freezes are free and guaranteed by federal law. Freeze your credit at all three bureaus:

- Equifax: equifax.com/personal/credit-report-services/credit-freeze
- Experian: experian.com/freeze/center.html
- TransUnion: transunion.com/credit-freeze

A freeze prevents anyone (including you) from opening new credit accounts. You can temporarily lift the freeze when you need to apply for loans, credit cards, or services requiring credit checks.

This doesn't stop identity thieves from accessing existing accounts, but it prevents them from opening new ones in your name, the attack that devastated Melissa.

Also, freeze credit for any children. Child identity theft often goes undetected for years because children don't check their credit reports. Criminals open accounts, run up debt, and disappear, leaving the child to discover ruined credit when they turn 18 and try to get their first credit card.

Opt Out of Pre-Approved Credit Offers

Visit optoutprescreen.com (operated by credit bureaus) to stop pre-approved credit and insurance offers. These offers don't just create mail clutter; they provide criminals with opportunities for mail theft and account opening.

You can opt out for five years or permanently. Permanent opt-out requires mailing a signed form, but it's worth the effort if you've experienced identity theft or live in an area with frequent mail theft.

Reality Check on Legal Rights

These rights sound powerful on paper. In practice, they face significant limitations:

Companies interpret exceptions broadly. They claim data is "necessary for business operations" or "required by law" to avoid deletion. They argue they've "anonymized" data, so privacy rights don't apply even though research proves anonymization often fails.

Enforcement remains weak. Regulatory agencies are understaffed and overwhelmed. Most privacy violations go uninvestigated.

Companies calculate that non-compliance costs less than compliance.

You bear the burden of proof. You must identify which companies have your data, submit properly formatted requests, follow up on non-responses, and potentially file complaints when companies ignore you. This takes hours for each request.

But imperfect rights exercised create more friction than no rights at all. When enough people submit deletion requests, companies must allocate resources to processing them, resources not spent on data collection. When enough people opt out of data sales, the product becomes less valuable. Individual actions compound into collective pressure.

Step 3: Reduce Attack Surface (Make Yourself Harder to Exploit)

You cannot eliminate all vulnerabilities, but you can reduce the number of entry points attackers can exploit.

Password Manager + Unique Passwords

Credential stuffing using breached passwords to access other accounts succeeded against Melissa because she reused passwords across sites. One breach exposed credentials that criminals used to access her bank, email, and eventually her entire digital life.

Password managers (1Password, Bitwarden, Dashlane, NordPass) generate and store unique passwords for every account. You remember one strong master password; the manager remembers everything else.

This prevents credential stuffing entirely. Even if attackers compromise your password on one site, they gain nothing useful for accessing other accounts.

Set up your password manager properly:

- Use a strong, unique master password (4-5 random words, at least 20 characters)
- Enable two-factor authentication on the password manager itself
- Store recovery codes in a secure location separate from your computer
- Start by changing passwords on critical accounts (email, banking, primary social media)
- Gradually migrate remaining accounts over time

Two-Factor Authentication (2FA) on Critical Accounts

2FA requires something you know (password) plus something you have (phone, security key, or authentication app code). Even if attackers steal your password, they cannot access your account without the second factor.

Enable 2FA on:

- All email accounts (especially the primary email)
- Banking and financial accounts
- Social media accounts
- Cloud storage (Google Drive, Dropbox, iCloud)
- Password manager
- Any account with payment information stored

Prefer authentication apps (Google Authenticator, Authy, Microsoft Authenticator) over SMS when possible. SMS can be intercepted through SIM-swapping attacks criminals convince your phone carrier to transfer your number to a device they control.

Security keys (YubiKey, Titan) provide the strongest protection but require purchasing hardware and managing physical devices. Use them for the highest-value accounts if you're willing to invest in additional security.

Review App Permissions Quarterly

Apps request access to contacts, location, camera, microphone, and other device features. Many request more permissions than they need for core functionality.

- Every three months:
- Review apps on your phone and computer
- Delete apps you haven't used in 30+ days
- Check permissions for remaining apps
- Revoke permissions that aren't essential for the app's function

Why does a flashlight app need your location? Why does a game need access to contacts? Why does a shopping app need microphone access? If you can't identify a legitimate reason, revoke the permission.

Go through connected services on major accounts:

- Google: myaccount.google.com/permissions
- Facebook: facebook.com/settings?tab=applications
- Apple: Settings > Privacy on iOS devices
- Microsoft: account.microsoft.com/privacy/app-permissions

Remove third-party apps and services you don't actively use. Each connected service creates a potential entry point if that service gets breached.

Email Hygiene

Create separate email addresses for different purposes:

- Primary personal email (friends, family, important accounts)
- Shopping/commercial email (retailers, newsletters)

- Throwaway email (one-time signups, untrusted sites)

This compartmentalization limits damage when addresses get compromised or sold to spammers. Criminals who breach a shopping site get your throwaway email, not the account connected to banking and critical services.

Services like SimpleLogin, DuckDuckGo Email Protection, or Apple's Hide My Email create unique forwarding addresses for each site. If a site sells your email or gets breached, you can disable that specific alias without changing your actual email address.

What This Achieves

These tactical protections make you a harder target than the majority who:

- Use the same password across multiple sites
- Never enable two-factor authentication
- Grant every permission app requests
- Use one email address for everything
- Never review or revoke access from unused services

Attackers are rational actors seeking maximum return for minimum effort. They target easy victims, people using "password123" across accounts, clicking every link in every email, leaving default settings unchanged.

You don't need perfect security. You need better security than the next person. That's what these strategies provide.

SECTION 3: WHY INDIVIDUAL ACTION ISN'T ENOUGH

The strategies in Section 2 matter. They reduce your vulnerability. They exercise what limited rights exist. They create friction for the systems profiting from your exposure.

But they cannot solve the problem that created Sarah's nightmare, Maria's insurance discrimination, Melissa's eighteen-month ordeal, Robert's financial devastation, or Lisa's algorithmic pregnancy detection.

Individual action fails against systemic threats because the Exposure Economy operates at a scale and speed that individual protection cannot match.

The Asymmetry Problem

When you protect one account with a strong password and two-factor authentication, you've secured one entry point. Attackers target millions of accounts simultaneously, using automated tools that test thousands of credentials per second. They only need one success to profit.

When you delete your file from one data broker, hundreds of other brokers still maintain profiles on you. New brokers emerge constantly, scraping data from sources you'll never identify. The ecosystem regenerates faster than you can dismantle it.

When you opt out of one tracking system, thousands more continue monitoring your behavior across websites, apps, and physical locations.

The digital advertising industry alone operates over 5,000 companies collecting and trading behavioral data. You cannot opt out of 5,000 systems individually.

This isn't a fair fight. You're one person with limited time, attention, and resources. The Exposure Economy is a $200 billion industry with unlimited automation, legal teams that specialize in avoiding privacy obligations, and business models that depend on your continued exposure.

The Collective Action Requirement

Privacy scholars increasingly frame privacy as a public health issue rather than an individual choice problem. This framing matters because it changes how we think about solutions.

Individual health decisions eating vegetables, exercising regularly, and avoiding cigarettes, improve your personal outcomes. But they don't solve public health crises caused by contaminated water, air pollution, or infectious disease. Those require collective action: regulations, infrastructure investment, coordinated public health campaigns.

Privacy works the same way. Your individual protections create modest personal benefit while leaving the exploitative infrastructure intact. The data brokers Melissa's attackers used continue operating profitably. The AI systems that predicted Lisa's pregnancy keep learning from millions of other users. The insurance companies that discriminated against Maria still purchase lifestyle analysis on new customers.

You protecting yourself doesn't weaken the system. It just shifts exploitation to people with less knowledge, fewer resources, or different vulnerabilities.

Real change requires addressing root causes through collective pressure:

Regulatory Action

Comprehensive federal privacy legislation could establish baseline protections for all Americans:

- Mandatory opt-in consent for data collection (not opt-out buried in terms of service)
- Right to access, deletion, and portability across all industries
- Prohibition on discriminatory algorithmic decision-making
- Private right of action allowing individuals to sue companies for violations.
- Meaningful penalties that exceed profits from non-compliance

The European Union's General Data Protection Regulation (GDPR) demonstrates that strong privacy law can change corporate behavior. After GDPR's 2018 implementation, companies faced billions in fines for violations. Many changed practices globally rather than maintain separate systems for European and non-European users.

The United States lacks equivalent federal protection. State laws like California's CCPA provide limited rights, but patchwork state regulations create compliance loopholes and leave residents of states without privacy laws completely unprotected.

Litigation and Enforcement

Class action lawsuits create accountability when regulatory enforcement fails. Recent settlements have forced changes in corporate data practices:

- Equifax paid $700 million after its 2017 breach exposed 147 million Americans (though individual victims like Sarah received minimal compensation)
- Facebook paid $650 million to settle Illinois biometric privacy claims

- Google, Apple, and other tech giants have faced antitrust cases targeting data monopolies.

These lawsuits don't make victims whole. Sarah's $6.50 settlement didn't compensate for her identity theft ordeal. But they impose costs that sometimes exceed profits from violations, changing corporate calculus about privacy protection.

The challenge: most terms of service include forced arbitration clauses preventing class actions. Regulatory action would need to ban these clauses in privacy contexts, giving consumers meaningful ability to hold companies accountable through courts.

Market Pressure

Consumer demand for privacy-protective alternatives can shift market dynamics. Signal and Telegram gained users when WhatsApp changed its privacy policies.

DuckDuckGo and Brave browsers grew as alternatives to Chrome's extensive tracking. Apple positioned privacy as a competitive advantage, though critics note gaps between marketing and practice.

But market solutions face inherent limitations. Privacy-protective services often cost money (people pay for what free services extract through data). This creates a "privacy divide" where wealthy people can buy protection while low-income people bear disproportionate surveillance and exploitation.

Relying on market competition also assumes consumers can make informed choices. But information asymmetry companies know everything about you while you know nothing about their practices, prevents genuine informed consent. You cannot meaningfully choose between services when you cannot evaluate their actual privacy practices.

What Collective Action Looks Like: Support Privacy Advocacy Organizations

Organizations fighting for systemic privacy protection include:

- **Electronic Frontier Foundation (EFF)**: Litigates privacy cases, advocates for privacy legislation, provides resources for individuals (eff.org)
- **Electronic Privacy Information Center (EPIC)**: Researches privacy issues, files policy complaints, publishes investigations (epic.org)
- **American Civil Liberties Union (ACLU)**: Challenges government surveillance and corporate privacy violations (aclu.org)
- **Consumer Reports**: Tests products for privacy risks, advocates for consumer privacy rights (consumerreports.org)

Membership, donations, and participation amplify these organizations' influence. They need resources to match the lobbying power of the industries profiting from the Exposure Economy.

Participate in Class Action Lawsuits

When you receive class action notices (often by mail or email), read them. Determine if you're affected by the breach or violation. Submit claims even when compensation is small.

Large numbers of claimants demonstrate widespread harm, strengthening the case for regulatory intervention. Even if you receive minimal individual compensation, participation contributes to collective accountability.

Monitor ongoing cases at classaction.org or through legal news sources. Major privacy class actions often take years to resolve, but create precedents affecting millions.

Contact Elected Representatives

Federal privacy legislation requires congressional action. Contact your senators and representatives:

- Express support for comprehensive privacy legislation
- Share personal stories of privacy violations (your experiences or those of family members)
- Request information about their positions on data protection
- Vote for candidates who prioritize privacy in their platforms

State-level advocacy matters too. Even in states with existing privacy laws, continued pressure strengthens enforcement and closes loopholes. In states without privacy laws, advocacy can drive new legislation.

Find your representatives at usa.gov/elected-officials. Organizations like EFF and EPIC provide template letters and specific legislative priorities.

Spread Awareness

Most people don't understand the Exposure Economy's scope until they experience a violation directly, like Sarah discovering her assembled profile after the attack, or Maria learning about lifestyle analysis only through her lawsuit.

- Sharing knowledge breaks information asymmetry:
- Recommend this book or similar resources to friends and family
- Explain privacy risks in accessible terms (not technical jargon that alienates people)

- Help others implement basic protections (show a family member how to freeze credit, set up a password manager, review app permissions)
- Call out privacy violations publicly when companies mislead users
- Cultural change precedes political change. When enough people recognize surveillance capitalism as exploitation rather than inevitable technology progress, democratic pressure for regulation becomes politically viable.

The Long Game

Collective action produces results slowly. Privacy legislation takes years from proposal to passage to enforcement. Class actions drag through courts for half a decade. Cultural shifts require generational change.

This timeline frustrates people experiencing urgent threats. If you're currently fighting identity theft like Melissa, you need solutions now, not promises of regulatory reform in 2030.

That's why individual and collective strategies must work together:

Individual action reduces your immediate risk while collective change develops. It's harm reduction for the present.

Collective action addresses root causes that individual protection cannot reach. It's a structural change for the future.

Neither approach alone succeeds. But combined, they create a path toward reclaiming digital identity at both personal and systemic levels.

You implement the strategies from Section 2 while supporting organizations fighting for the changes in this section. You protect yourself today while building power for tomorrow.

The Exposure Economy won't dismantle itself. It requires deliberate, sustained, collective resistance.

SECTION 4: PSYCHOLOGICAL RECLAMATION

The hardest form of reclamation has nothing to do with technology.

Lisa can delete her pregnancy app, opt out of data brokers, and implement every technical protection available. But the algorithm's prediction its knowledge of her pregnancy before she consciously recognized it changed something fundamental. She cannot unknow that artificial intelligence understands her body's patterns better than she does.

Maria won her insurance lawsuit. But she cannot erase the knowledge that companies are making predictions about her health, finances, and behavior that she's not allowed to see or challenge. Those predictions shape opportunities and obstacles in ways she'll never fully understand.

Melissa proved the fraudulent accounts weren't hers. But employers still ask her to explain why background checks show identity theft. She carries proof of innocence everywhere, defending herself against crimes committed by strangers using her stolen identity.

The deepest violation isn't data theft. It's identity theft in a more profound sense, having your actual self-replaced by an algorithmic approximation, then being judged, priced, and limited by that approximation regardless of its accuracy.

Psychological reclamation means refusing to let algorithmic predictions define who you are and what you're allowed to become.

Understanding How AI Learned to Manipulate You

The research from 2013 on gaming psychology revealed something game designers had discovered through experimentation: hedonic factors, enjoyment, self-concept, self-efficacy, and self-congruity drive behavior more powerfully than utilitarian factors like ease of use or objective value.

Gamers didn't buy games to accomplish tasks. They bought games to feel a certain way about themselves. The achievements, scores, and progression were just vehicles for emotional rewards.

AI systems trained on gaming data learned this lesson perfectly. Then they applied it to everything else.

When you see an ad that makes you feel inadequate unless you buy the product, that's self-concept exploitation. The algorithm knows you aspire to a certain identity and positions the product as the bridge between your actual self and your ideal self.

When you experience FOMO (fear of missing out) from social media posts showing friends at events you weren't invited to, that's self-congruity manipulation. The algorithm knows you value belonging and shows you content designed to trigger anxiety about social exclusion.

When you feel compelled to click a notification even though you know it's probably unimportant, that's variable reward schedule exploitation, the same mechanism that makes slot machines addictive. The algorithm knows that unpredictable rewards drive compulsive behavior more effectively than consistent rewards.

When you "doom scroll" through disturbing news despite wanting to stop, that's emotion-driven engagement optimization. The algorithm knows that anger, fear, and outrage hold attention longer than positive content, so it feeds you material designed to trigger those responses.

These aren't accidents. They're deliberate applications of psychological insights originally developed to maximize game purchases and usage, now deployed across every domain of digital life.

Recognizing Hedonic Manipulation in Real Time

Psychological reclamation starts with awareness. You cannot resist manipulation you don't recognize.

The Urgency Test

When you feel a sudden compulsion to buy, click, share, or respond, pause. Ask: "Where did this urgency come from?"

Authentic needs rarely create artificial time pressure. "Limited time offer!" and "Only three left in stock!" are manufactured scarcity designed to bypass rational evaluation. If you need the product, it will still be worth buying tomorrow.

Real urgency comes from external circumstances: your car broke down, you ran out of necessary medication, or you need to replace a broken appliance. Digital urgency almost always comes from manipulation.

The Identity Test

When marketing positions a product as essential to who you are or who you want to become, stop. Ask: "Am I responding to authentic self-expression or manufactured desire?"

Authentic identity alignment happens when products support existing values and practices. If you already run daily, running shoes serve real needs. If you're vaguely thinking you should exercise more someday, the shoes are an identity aspiration, buying the appearance of being a runner rather than running.

AI excels at identity manipulation because it knows the gap between your actual self and your ideal self. It has months or years of behavioral data showing what you do versus what you think you should do. It targets that gap deliberately.

The Social Proof Test

When you see "trending," "most popular," or testimonials from people like you, pause. Ask: "Am I evaluating this on its merits or just following the crowd?"

Social proof exploits your need for belonging and fear of missing out. "Join 10 million users" doesn't tell you whether the product is actually good, just that many people use it, possibly for reasons that don't apply to you.

AI-generated social proof is often fake or strategically curated. "People like you also bought..." might mean "we've segmented you into a demographic category and these are the products with the highest profit margins for that segment."

The Emotion Test

When content makes you feel strong emotions, rage, fear, excitement, or sadness before you fully understand what you're reacting to, stop. Ask: "Is this designed to make me feel this way?"

Emotional manipulation works because it bypasses rational evaluation. You click, share, or comment based on immediate feeling before conscious thought catches up.

News designed to maximize engagement emphasizes conflict, danger, and outrage. Social media algorithms amplify emotionally charged content because it drives more interaction. Even if the information is technically accurate, the presentation is engineered for emotional impact rather than understanding.

Reclaiming Self-Concept from Algorithmic Approximation

Your actual identity exists independently of your digital profile. This sounds obvious, but the Exposure Economy works hard to make you forget it.

Every time an algorithm denies you an opportunity, you're experiencing judgment based on an approximation. The AI doesn't know you. It knows patterns derived from millions of people who share some characteristics with you, patterns that may not apply to you individually.

When Maria's insurance premium increased due to "lifestyle analysis," the algorithm made predictions based on statistical correlations. It didn't know that Maria runs five days weekly, doesn't smoke, and has perfect health markers. The prediction mattered more than the reality.

Document Where Predictions Are Wrong

Keep records when algorithmic decisions misrepresent you:

- Loan denials despite strong credit and income
- Job rejections after automated resume screening
- Insurance rate increases without claims or health changes
- Product recommendations are completely inconsistent with your interests
- Content feeds showing material you actively dislike

This documentation serves two purposes. First, it reminds you that the algorithm is often guessing incorrectly. Second, it provides evidence when you challenge decisions or participate in discrimination lawsuits.

Demand Human Review

When automated systems make significant decisions about you, request human review. Many institutions are required to provide it, but don't advertise that option.

For credit decisions, lenders must provide "adverse action notices" explaining denials and your right to request human review. For employment, automated resume screening is increasingly restricted by state laws requiring human involvement. For insurance, many states mandate review processes for rate increases.

Human reviewers still rely partly on algorithmic outputs, but they can consider context that the AI cannot process. Explain discrepancies. Provide documentation contradicting the profile. Appeal to judgment rather than just pattern matching.

Actively Curate Counter-Narratives

If algorithms categorize you in ways you reject, deliberately create data contradicting those profiles.

This doesn't mean lying or fabricating information. It means being intentional about which aspects of your multifaceted identity you emphasize in different contexts.

If algorithms assume you're financially struggling based on zip code and shopping patterns, but you're actually financially secure, make purchases at retailers and use services associated with your actual economic status. The algorithm will gradually update its categorization.

If you're categorized based on age-related assumptions that don't apply to you, engage with content and products that contradict those stereotypes.

Over time, personalization systems adjust, though this creates the perverse burden of performing your identity for algorithmic audiences.

Digital Autonomy Practices

Reclamation ultimately means recovering agency, making conscious choices rather than reactive responses.

Deliberate vs. Reactive Technology Use

Most people use technology reactively:

- Notifications tell you what to look at
- Recommendations tell you what to watch
- Feeds tell you what to read
- Ads tell you what to buy
- Algorithms tell you what to care about

Your behavior is a response to prompts designed to maximize engagement and extract value. You're not choosing, you're reacting to choices made by systems optimizing for corporate profit.

Deliberate technology use reverses this relationship:

- Decide what you want to accomplish
- Choose the appropriate tool
- Use it for that specific purpose.
- Close it when done

Instead of opening social media and scrolling through whatever appears in your feed, decide, "I want to check if my sister posted about her vacation." Open the app, look at her profile, close the app. Don't react to the algorithm's suggestions about what else you should see.

Instead of watching whatever Netflix recommends, decide "I want to watch a specific movie I've been meaning to see." Search for it, watch it, turn off the TV. Don't autoplay into algorithmic recommendations.

This sounds simple. It's extraordinarily difficult because interfaces are designed to prevent deliberate use. Infinite scroll, autoplay, notification badges, and recommendation engines all push you toward reactive consumption.

Building Autonomy Gradually

Start small. Choose one technology interaction daily where you'll practice deliberate use:

- Set a timer before opening social media (10 minutes to check messages, then close)
- Turn off autoplay on video platforms
- Disable notification badges so you check apps on your schedule, not the app's schedule
- Use website blockers during focus time
- Keep your phone in another room during meals or family time

These practices feel restrictive initially because you've adapted to reactive patterns. The algorithm has trained you to expect constant stimulation, immediate response, and endless novelty.

Reclaiming autonomy means retraining yourself to tolerate boredom, delay gratification, and choose deliberately rather than consume reactively. It's harder than any technical privacy protection because it requires changing habits reinforced by systems designed to make those habits compulsive.

Accepting Imperfection

You will fail at this repeatedly. You'll intend to check one thing and find yourself thirty minutes later deep in algorithmic recommendations. You'll disable notifications and then re-enable them because you fear missing something important.

You'll commit to deliberate use and slip back into reactive patterns.

This is normal. You're resisting systems optimized over billions of user interactions to maximize addictive behavior. They're good at what they do.

Progress isn't perfection. It's increasing the percentage of interactions you control rather than react to. It's catching yourself mid-scroll and closing the app instead of continuing for another hour. It's noticing manipulation attempts even when you can't always resist them.

Psychological reclamation is the longest game. Technical protections take hours to implement. Legal rights take weeks to exercise. Collective action takes years to produce change.

Reclaiming your mind from systems designed to control it takes constant, ongoing effort for as long as you participate in digital life.

But it's the only form of reclamation that's truly yours. Data brokers can ignore deletion requests. Companies can lobby against privacy laws. But no algorithm can force you to believe its predictions about who you are and what you should become.

Your actual identity, the one you know through lived experience rather than statistical inference, remains yours to define. Protecting that requires technical and legal strategies.

But reclaiming it requires refusing to let approximations replace reality, regardless of how sophisticated the algorithms become.

SECTION 5: WHAT HAPPENS NEXT

Reclamation isn't a destination. It's an ongoing practice that adapts as technology evolves, threats change, and your circumstances shift.

Understanding realistic timelines for different types of reclamation helps set appropriate expectations and avoid the frustration that leads people to give up entirely.

Short-Term Reclamation (30-90 Days) What You Can Accomplish:

Within three months, you can implement most individual protection strategies from Section 2:

- Complete exposure audit (data breach checks, broker requests, social media review)
- Set up a password manager and migrate 50+ critical accounts
- Enable two-factor authentication on essential services
- Submit deletion requests to major data brokers
- Freeze credit for yourself and family members
- Review and revoke unnecessary app permissions
- Establish basic deliberate technology use practices

Expected Outcomes:

You will not feel dramatically safer. The Exposure Economy doesn't shrink noticeably. But you will have:

- Reduced your attack surface (credential stuffing now fails against you)
- Exercise existing rights (even if compliance is partial)
- Begun building awareness of manipulation attempts
- Created a foundation for longer-term practices

Likely Frustrations:

About 43% of data broker deletion requests will be ignored. You'll discover new people-search sites listing your information faster than you can remove old listings. You'll catch yourself doom-scrolling despite commitments to deliberate use. Two-factor authentication will occasionally feel inconvenient.

This is normal. Persistence matters more than perfection.

What to Measure:

Don't measure "am I safe?" (answer is always no in absolute terms). Measure:

- How many accounts now have unique passwords + 2FA?
- How many data broker files have you accessed or requested deletion for?
- How many times this week did you catch yourself about to react to manipulation and choose differently?

These concrete metrics show progress even when threats persist.

Medium-Term Reclamation (1-3 Years) What Requires Sustained Effort:

Some reclamation strategies show results only over extended periods:

Legal and Regulatory Action: Privacy legislation moves slowly. The California Consumer Privacy Act took years from proposal (2018) to implementation (2020) to enforcement (ongoing). Federal privacy legislation has been proposed repeatedly since 2019 without passage.

Class action lawsuits average 2-5 years from filing to settlement. The Equifax breach happened in 2017; the settlement wasn't finalized until 2022.

Your participation in contacting representatives, joining advocacy organizations, and filing complaints with regulatory agencies contributes to pressure that takes years to produce results.

Algorithmic Profile Adjustment: If you're actively curating counter-narratives to challenge incorrect algorithmic categorizations, AI systems update gradually through exposure to new behavioral data. A few deliberate purchases won't immediately reclassify you. Consistent patterns over months or years shift how algorithms predict your behavior.

This timeline frustrates people seeking immediate results, but it reflects how machine learning works. Models trained on years of data require substantial contradictory evidence to change predictions.

Cultural Shift Around Privacy When you help family and friends understand privacy threats and implement protections, cultural change spreads through networks slowly. One conversation rarely converts someone immediately. But repeated exposure to privacy concerns, personal stories, and accessible strategies gradually shifts norms.

Ten years ago, most people didn't use password managers or understand data brokers. Today, awareness is growing, still inadequate but improving through exactly this kind of network diffusion.

Expected Outcomes:

Within 1-3 years of consistent practice:

- Privacy protections become habitual rather than requiring constant conscious effort
- You've participated in multiple rounds of advocacy (legislative sessions, commenting periods, class actions)
- Your social circle shows increased privacy awareness and protective behaviors
- You've experienced at least one instance where protections prevented harm you would have experienced otherwise

- Algorithmic personalization targeting you shows some evidence of adjustment

What to Measure:

- Are privacy practices now automatic habits or still a conscious effort?
- How many people in your network have you helped implement protections?
- What privacy legislation has advanced (even if not yet passed)?
- Have you received fewer spam calls/emails after deletion requests and email hygiene?

Long-Term Reclamation (3-10 Years) Systemic Change Timeline:

Fundamental restructuring of surveillance capitalism operates on generational timelines.

The civil rights movement took decades from initial organizing to legislative success to cultural transformation. Environmental regulation took twenty years, from early advocacy to the establishment of the EPA to the implementation of significant protections.

Privacy regulation follows similar patterns. Europe's journey from recognizing privacy as a fundamental right (1950s) to implementing GDPR (2018) spanned sixty years. The United States is earlier in that trajectory.

What Long-Term Participation Enables:

Sustained engagement over the years:

- Builds organizational power (advocacy groups grow stronger with consistent member support)
- Creates political viability (when enough constituents prioritize privacy, politicians respond)
- Establishes legal precedents (early lawsuits shape interpretation for decades)
- Shifts cultural expectations (privacy becomes assumed rather than exceptional)

Realistic Endpoints:
Even with sustained collective action, you will not achieve:

- Complete elimination of data brokers (though meaningful regulation is possible)
- Total opt-out from surveillance capitalism while participating in digital society
- Perfect privacy (always an unattainable goal). You can contribute to:
- Comprehensive federal privacy legislation with enforcement mechanisms
- Meaningful penalties for violations that change corporate cost-benefit calculations
- Cultural norms that treat privacy invasion as unacceptable rather than inevitable
- Youth growing up with privacy literacy rather than learning through violation

What to Measure:
Progress over the decades has shown in:

- Legislative wins (bills passed, regulations implemented, enforcement actions taken)
- Court precedents (cases establishing privacy rights and corporate accountability)
- Market shifts (privacy-protective products becoming mainstream rather than niche)
- Generational change (younger people approaching privacy differently than their parents)

The Realistic Endpoint

Sarah still feels digitally homeless. Maria's profile still exists in hundreds of databases. Melissa still explains her identity theft to background check reviewers. Robert has withdrawn from digital life almost entirely. Lisa's daughter is being profiled from birth.

None of them achieved complete reclamation. That's not possible in the current system.

But each could:

- Reduce ongoing exploitation through individual protections
- Exercise legal rights to delete accessible data
- Resist psychological manipulation more effectively over time
- Contribute to collective efforts to change the system
- Live with technology on slightly more equitable terms

That's not the victory we want. It's the victory available given the present reality.

Reclamation means refusing to accept the present reality as permanent. It means working simultaneously on individual protection (immediate harm reduction) and systemic change (long-term structural reform).

You implement password managers today while supporting privacy legislation for tomorrow. You freeze your credit now while joining class actions that take years to resolve. You practice deliberate technology use this week while contributing to cultural shifts that take decades.

The Exposure Economy won't dismantle itself. But it's not invulnerable either.

Every deletion request honored is friction in the system. Every person who learns to recognize manipulation is resistance. Every privacy law passed is progress.

Every company that faces meaningful penalties is accountability.

Reclamation is possible not as individual restoration to pre-digital privacy, but as a collective movement toward a different relationship between people and the systems that profile them.

The question isn't "Can I reclaim my digital identity completely?"

The question is "What can I reclaim today, this month, this year, and what systemic change can I contribute to for people who come after me?"

Special Considerations: Protecting Family Identity

Lisa's daughter didn't choose to have her behavior tracked from birth. Data brokers began assembling files on her before she could talk, based on her parents' purchases, location patterns, and online searches.

If you have children, your reclamation practices must extend to protecting identities that they cannot yet protect themselves.

The Family Exposure Challenge

Children face unique vulnerabilities:

- They cannot consent to data collection, but live in environments saturated with tracking
- Their identity formation happens under constant surveillance; they don't understand
- Mistakes they make online create permanent records, affecting future opportunities
- Information parents share about them builds profiles children inherit without choice

Research shows 82% of parents post photos of their children online, but only 18% consider long-term privacy implications. Most share their child's photo, name, and birthday together without realizing this combination enables identity theft.

By age two, the average child has 1,000 photos posted online by parents and relatives. By age five, they appear in 1,500+ posts, creating detailed behavioral timelines. Before they're old enough to understand privacy, their digital identity is already constructed.

Age-Appropriate Protection Strategies Young Children (Ages 5-12)

Children this age need protection they cannot provide themselves:

What Parents Control:

- Never post full names, birthdates, and locations together
- Avoid location tags on current activities (tag after you've left)
- Crop or blur identifying information in photos (school names, house numbers, license plates)

- Share in private groups rather than public posts
- Ask other parents before posting photos with their children

What Children Learn:

- "Trusted adult rule": Always ask a trusted adult before sharing personal information online
- Strangers online are like strangers in real life; friendly doesn't mean safe
- Some information (full name, address, phone number, school name) is never shared publicly

Use simple analogies: passwords are like house keys; personal information is like your home address, you don't give it to people you don't know.

Teens (Ages 13-18)

Teenagers need privacy to develop independence, but also protection from poor digital decisions that create permanent records.

Research shows "power imbalances between parents and teens hinder collaboration and lead to resistance or secrecy." Heavy-handed control backfires. Teens find ways around restrictions and stop communicating with parents about online problems.

What Works Better:

Collaborative privacy management:

- Negotiate social media rules together rather than imposing them unilaterally
- Focus on consequences ("this post could affect college admissions") rather than just rules
- Practice the "grandmother test": Would you be comfortable if your grandmother saw this?
- Review privacy settings together quarterly, explaining why each matters

What Fails:

Secret monitoring destroys trust faster than it prevents harm. If you monitor, be transparent about what and why. Gradually reduce monitoring as teens demonstrate responsible judgment.

Protecting Children's Financial Identity

Child identity theft often goes undetected for years. Criminals open credit accounts in children's names, run up debt, and disappear, leaving the child to discover ruined credit at 18 when they try to get their first credit card or student loan.

Immediate Actions:

Freeze your children's credit at all three bureaus. The process requires:

- Sending copies of your ID and the child's birth certificate
- Proof of address
- Written request to each bureau

It's inconvenient, but it prevents criminals from opening

accounts. Unfreeze temporarily if your child needs to establish credit as a young adult.

Check if your child has existing credit files. Contact bureaus and request reports. If files exist when they shouldn't, investigate immediately it may indicate identity theft has already occurred.

The "Privacy-First Sharing" Approach

You don't have to choose between sharing family moments and protecting children. You need strategies that allow both:

Before Posting:

- Ask yourself: "Will my child want this online when they're applying to college/jobs?"
- Crop identifying information (blur school logos, street signs, house numbers)
- Remove metadata (location data embedded in photo files)
- Consider who needs to see this

Where to Share:

- Private family groups instead of public posts
- Encrypted messaging apps rather than social media
- Photo-sharing services with strict access controls
- Physical photo albums for content too sensitive for any digital sharing

What Never to Share:

- Bathroom/bathing photos
- Discipline or meltdowns (embarrassment that follows children for years)
- Medical information or diagnoses
- Information about difficult family situations

Device and Account Security For Children's Devices:

- Set up parental controls thoughtfully (they have security vulnerabilities, keep software updated)
- Create strong passwords children can remember (passphrases work well)
- Enable two-factor authentication on app stores (prevents unauthorized purchases and malware installation)
- Set age-appropriate screen time limits
- Review installed apps monthly and remove those unused

For Shared Family Devices:

- Create separate user accounts (don't share logins)
- Use a password manager for family accounts (shared access to Netflix/Spotify credentials, not personal accounts)
- Secure home network (change router default password, use WPA3 encryption)
- Set up a guest network for visitors and smart home devices

When Protection Fails: Family Emergency Response

Despite protections, breaches happen. Have family emergency procedures:

If a Child's Information Is Compromised:

- Freeze credit immediately if not already frozen
- Document everything (dates, communications, fraudulent accounts)
- File reports with FTC (identitytheft.gov) and local police
- Contact credit bureaus to flag the child's file
- Monitor regularly until the child reaches adulthood

If a Parent's Breach Affects Family:

- Assess what information could be exposed that could identify children
- Change compromised passwords that children might have observed
- Discuss with age-appropriate children what happened and what changed
- Review and tighten family privacy rules based on lessons learned

Building Long-Term Privacy Literacy

The goal isn't protecting children forever; it's teaching them to protect themselves.

Age 5-8: Basic concepts (strangers, trusted adults, public vs. private),

Age 9-12: Understanding consequences (permanence of online posts, identity theft concepts).

Age 13-15: Practicing judgment (social media decision-making with guidance).

Age 16-18: Independence with oversight (making privacy decisions while discussing reasoning),

By 18, children should be able to:

- Recognize phishing and social engineering attempts
- Manage their own passwords and two-factor authentication
- Make informed decisions about what to share online
- Understand their legal privacy rights
- Implement basic protection strategies independently.

This requires a gradual transfer of control, not a sudden withdrawal of all protection at legal adulthood.

The Hard Balance

Over-protection prevents children from learning digital responsibility. If parents control everything, children never develop their own privacy skills and become vulnerable when they gain independence.

Under-protection leaves children unprepared for real threats. Without education and tools, they make dangerous mistakes that could have been prevented.

The balance: Age-appropriate protection that decreases as competence increases, combined with education that builds genuine understanding rather than just rule-following.

Family privacy reclamation means protecting the identities of people who cannot yet protect themselves while building their capacity to eventually take over that protection.

Your children will inherit the Exposure Economy you're fighting to change. What they inherit depends partly on the collective action discussed in Section 3.

But it also depends on whether you protect their identity in the present while teaching them to protect it themselves in the future.

The journey of reclamation, individual, collective, psychological, and familial, begins with a single decision: refusing to accept that the Exposure Economy's current form is inevitable.

Sarah, Maria, Melissa, Robert, and Lisa cannot undo what happened to them. But their experiences reveal how the system works and where it's vulnerable to resistance.

Your reclamation strengthens theirs. Every person who understands how identity gets extracted and weaponized is one more person who can resist, advocate, and contribute to changing the system that made their violations possible.

CONCLUSION

Reclamation is not restoration. You cannot return to a state of pre-digital privacy that never truly existed. Data collected will not vanish. Algorithms trained on your behavior will continue making predictions. The infrastructure of the Exposure Economy, the $200 billion industry built on constant visibility, persists regardless of individual protective action.

But reclamation is recovery. Recovery of agency in a system designed to eliminate choice. Recovery of understanding in an economy that profits from your confusion. Recovery of power to shape how you participate in digital life rather than simply enduring extraction as the cost of participation.

The Exposure Economy operates on a foundational premise: your constant visibility creates value that corporations capture while you bear the risks.

Surveillance capitalism refined this model, claiming your private experience as free raw material for behavioral prediction and modification. Together, they constructed a world where exposure is mandatory, asymmetry is structural, and resistance seems futile.

But exposure isn't inevitable; it's enforced through design

choices, business models, and regulatory failures that benefit from your belief that nothing can change.

Asymmetry isn't natural; it's manufactured through deliberate information hiding, incomprehensible terms of service, and legal frameworks written by the industries they supposedly regulate. And resistance isn't futile, it's happening in every deletion request that gets honored, every person who learns to recognize manipulation, every law that imposes real penalties, every collective action that demands accountability.

Sarah, Maria, Melissa, Robert, and Lisa cannot undo what happened to them. But their experiences map the territory. They show where the Exposure Economy is vulnerable: in the gap between its promises and its practices, between its claims of inevitability and its actual fragility when challenged, between individual powerlessness and collective strength.

You now see what they saw too late. That sight is uncomfortable. It reveals how much has already been taken, how much continues to be extracted, and how the systems you depend on daily are designed to exploit rather than serve you.

But sight is also power. You cannot reclaim what you cannot see. You cannot resist what you don't recognize. You cannot change what you believe is unchangeable.

The Exposure Economy and surveillance capitalism are not natural laws. They are choices about how technology is designed, how data is regulated, how value is distributed, and how power is concentrated. Choices can be challenged. Systems built on choices can be rebuilt differently.

Individual reclamation reduces your exploitation. Psychological reclamation protects your autonomy. Collective reclamation changes the system. Family reclamation protects those who cannot yet protect themselves.

None alone is sufficient. Together, they form a path from exposure to agency, from surveillance to resistance, from endurance to power.

The Exposure Economy will not dismantle itself. Surveillance capitalism will not voluntarily abandon its business model. The systems profiting from asymmetry will not surrender their advantages without pressure.

But they are not invulnerable. Every person who refuses to accept total surveillance as the price of digital participation weakens the assumption that exposure must be mandatory. Every collective action challenging corporate extraction creates precedent for accountability. Every law recognizing privacy as a right rather than a consumer preference shifts the foundation on which the Exposure Economy is built.

Reclamation begins with one decision: refusing to accept the current system as permanent. What you do with that decision today, this month, this year determines not just your own exposure, but the world those who come after you will inherit.

The choice is not between perfect privacy and total surveillance. It is between accepting the Exposure Economy's terms or reclaiming enough power to renegotiate them.

Your reclamation starts now.

PRACTICAL CHECKLIST

Essential Privacy & Security Checklist (Streamlined)

Use this checklist to implement protections in manageable stages. Check boxes as you complete each action.

Foundation (Do This First - 30 minutes) Secure Critical Accounts

- [ ] Enable two-factor authentication (2FA) on email, banking, and primary social media
- [ ] Use an authentication app (Google Authenticator, Authy) instead of text messages
- [ ] Create unique passwords for each important account (minimum 12 characters)
- [ ] Set up the phone's built-in password manager OR download a dedicated password manager

Lock Down Devices

- [ ] Turn off location tracking for apps that don't need it
- [ ] Review app permissions (Settings > Privacy) - revoke unnecessary access
- [ ] Delete apps unused in the past 6 months
- [ ] Require password/PIN for app downloads and purchases

Check Your Exposure

- [ ] Google your full name in quotes - screenshot first page of results
- [ ] Check haveibeenpwned.com for known data breaches
- [ ] Search your phone number and email address online

Privacy Protection (Next 30 minutes) Social Media Cleanup

- [ ] Change all social media to private/friends-only
- [ ] Remove phone number from public profiles
- [ ] Turn off tagging without permission
- [ ] Delete posts showing full address, children's schools, vacation plans while traveling

Monitor Your Identity

- [ ] Freeze credit at all three bureaus (free at Equifax, Experian, TransUnion websites)
- [ ] Set up Google Alerts for your name, phone, and email
- [ ] Check free credit report at annualcreditreport.com

Advanced Protection (30-60 minutes) Limit Data Collection

- [ ] Turn off ad personalization: Google (myaccount.google.com/data-and-privacy), Facebook (Settings > Ads)
- [ ] Disable interest-based ads in phone settings
- [ ] Review which apps access your Google/Facebook account - remove unused ones
- [ ] Turn off voice recording on smart devices unless actively needed

Use Privacy Tools

- [ ] Install privacy-focused browser (Brave, Firefox) OR privacy extensions (uBlock Origin) on current browser
- [ ] Use private/incognito mode for sensitive searches
- [ ] Create separate email addresses: one for important accounts, one for shopping, one for throwaway signups

Family Protection (30-45 minutes if applicable) Protect Children

- [ ] Freeze children's credit (requires mailing documents to each bureau)
- [ ] Limit sharing: never post a child's full name + birthdate, + location together
- [ ] Review photos for identifying information (school names, house numbers, uniforms)
- [ ] Set up parental controls on family devices
- [ ] Teach children ages 10+: never share personal info without asking a trusted adult

Family Privacy Rules

- [ ] Hold a 15-minute family discussion about what not to share online
- [ ] Create rule: ask permission before posting photos of family members
- [ ] Establish verification procedure: if someone calls claiming an emergency, hang up and call them back at a known number

Ongoing Maintenance Monthly (15 minutes)

- [ ] Google yourself - check for new exposed information
- [ ] Review app permissions - revoke unnecessary access
- [ ] Update passwords on 3-5 important accounts
- [ ] Check credit report (free weekly at annualcreditreport.com)

Quarterly (30 minutes)

- [ ] Review and delete old social media posts
- [ ] Update privacy settings on all platforms
- [ ] Remove old apps and unused online accounts
- [ ] Test family scam awareness (practice recognizing phishing emails together)

Annually (1 hour)

- [ ] Download your data from Google (takeout.google.com), Facebook, and Amazon
- [ ] Review all online accounts - close ones you don't use
- [ ] Update emergency response plan and fraud hotline numbers

Emergency Response (If Identity Theft Occurs) Immediate (Within 24 hours)

1. Call banks/credit cards - report fraud
2. Freeze credit at all three bureaus if not already frozen
3. File a police report
4. Document everything with screenshots and notes

Follow-Up (Within 1 week)

1. File FTC report at IdentityTheft.gov
2. Review ALL financial accounts for unauthorized activity
3. Change passwords on all important accounts

Recovery (Ongoing)

1. Monitor credit reports monthly
2. Keep detailed records of all recovery steps
3. Follow up on all fraud reports until resolved

Quick Priority Guide (When Time Is Limited)

5 minutes: Enable 2FA on email + Turn off unnecessary app location tracking
15 minutes: Set up password manager + Make social media private
30 minutes: Freeze credit at one bureau + Review app permissions
1 hour: Complete all Foundation steps + Set up monitoring

ABOUT THE AUTHOR

Neil Singh bridges consumer psychology, virtual environments, and cybersecurity, a combination that positioned him to recognize threats most experts missed until it was too late.

His journey began unconventionally. While earning his master's degree at Unitec Institute of Technology in New Zealand, Singh studied why people buy and play video games. His research revealed something gaming companies had discovered through trial and error: emotional triggers, self-concept, belonging, and status drove purchasing decisions far more powerfully than rational factors like ease of use or price.

The 2013 publication of his research on the Utilitarian-Hedonic Dual Mediation Model documented how virtual environments exploit psychological vulnerabilities to maximize engagement and spending. At the time, this seemed relevant only to gaming. A decade later, those same manipulation techniques, now automated by AI, operate across every domain of digital life.

Singh's work in metaverse strategy and Industrial IoT with automotive, aerospace, and industrial clients revealed how quickly virtual environment psychology was scaling beyond entertainment. Organizations implementing digital twins and immersive customer experiences were unknowingly creating new attack surfaces while amplifying existing psychological vulnerabilities.

As AI capabilities accelerated, Singh shifted focus to the intersection where his expertise became most urgent: cybersecurity, privacy, and the governance challenges of deploying AI systems that learn from and exploit human behavior.

His 2025 presentation at Augmented World Expo on *"Artificial General Intelligence in Metaverse and Web3: Risks and Threats to National Security"* synthesized years of pattern recognition: AI didn't invent behavioral manipulation. It is industrialized techniques that gaming and advertising have refined over decades, and then deployed them at scale across finance, healthcare, insurance, employment, and democratic systems.

Singh's commitment to digital equity shapes his approach to privacy and security. Through workforce development initiatives that train underrepresented individuals for technology careers, he's seen firsthand how privacy violations disproportionately harm those with the fewest resources to fight back. Technical solutions alone cannot address systemic problems rooted in power asymmetry.

This book represents the convergence of Singh's academic research into consumer psychology, his practical experience implementing digital systems across industries, and his recognition that privacy and security require understanding human behavior as deeply as technical architecture.

The Exposure Economy isn't just a privacy problem or a security problem. It's a human problem that demands solutions grounded in how people think, decide, and trust, not how systems assume they should.

www.ingramcontent.com/pod-product-compliance
Lightning Source LLC
LaVergne TN
LVHW022102200926
848481LV00003B/1039